Data-Mining and Intermetallic Property-Prediction

David J. Fisher

Published by **Materials Research Forum LLC**
Millersville, PA 17551, USA

Published as part of the book series
Materials Research Foundations
Volume 128 (2022)
ISSN 2471-8890 (Print)
ISSN 2471-8904 (Online)

Print ISBN 978-1-64490-200-4
ePDF ISBN 978-1-64490-201-1

Distributed worldwide by

Materials Research Forum LLC
105 Springdale Lane
Millersville, PA 17551
USA
http://www.mrforum.com

Printed in the United States of America
10 9 8 7 6 5 4 3 2 1

Table of Contents

Introduction

The basic problem considered here is the prediction of the structure of an intermetallic alloy and of its associated properties and how this once seemingly intractable task has yielded to the computer-aided search for patterns among the vast databases of tabulated experimental results. Taking a truly global view it has been pointed out[1] that the entire material world, including known and unknown compounds, must lie on an energy landscape which includes the free energies of those compounds: that is, all possible chemical compounds are present in this landscape. Synthesis then corresponds to compound-discovery rather than compound-creation. Materials are, in effect, discovered during experimental exploration of the energy landscape. Given that vast databases now exist which represent large tracts of this energy landscape, it can be hoped that more rapid progress can be made by mining them for information and for spotting correlations which can then act as maps for guiding a more efficient exploration of the landscape. The intermetallic phases alone exhibit a wide variety of crystal structures, with unit-cell sizes ranging from 1 to more than 20000 atoms, and having many symmetries, stoichiometries, structures and stability fields. A total of 20829 structures, of 2166 types, was studied[2]. There was a sub-set of 6441 binary intermetallic compounds which exhibited 943 structural types.

It can be argued that all of science is a search for patterns in Nature and, once a pattern has been discovered, that then offers the possibility of prediction. The power of prediction subsequently leads to a great deal of time-saving as the scientist is now armed with a guide to where to look for new fruitful fields of study … and which dead-ends to avoid.

With regard to the latter, the mind immediately turns to Kepler and his claim that the planetary orbits corresponded to a 'nesting' of the Platonic solids[3]. Closely related to this is Bode's Law[4], which is notoriously prevalent and well-supported by data while frustratingly offering no agreed rationale for its existence. There is therefore always the danger that the observed pattern is illusory, and is the result of pure numerology and its mystical overtones; especially when π or the Golden Section is implicated. Gamow pointed out[5] many examples of this tendency, and there had been quite a fashion in the mainstream physics literature of the early decades of the 20th century to insert π into basic physical relationships[6] at every opportunity. Some scientific celebrities have fallen foul of forms of numerology; most notably Arthur Eddington[7]. Dirac's views on 'large number coincidences' meanwhile continue to command respect.

At the same time, it is necessary to be careful to distinguish between an honestly felt analogy which closely parallels the then-available data, and mere wishful thinking. The mathematician, Sylvester, surmised that purely mathematical entities explained the nature of the elements[8]. Tait had similarly suggested that the elements were various species of 'knots in space'.

The point of this digression is that some current mature and established branches of science began as what might well be deemed numerology. Crick memorably discussed[9] a 'magic number' theory in the early days of DNA research, and Svedberg had in fact fallen into the numerology-trap in his related work on proteins[10]. The concept of 'magic numbers' in protein research persisted[11], but with more dependable experimental underpinning. The concept of the magic number is also often to be found in other sciences. For example, there is the observation that certain small 'magic' groupings of atoms are particularly stable[12] and other 'magic' groups may be particularly liable to diffuse across substrates[13].

Undoubtedly the most important pattern in Nature ever discovered is the Periodic Table of the Elements, with its ability to guide all aspects of ongoing scientific research. The subject of the present work is how this aspect is now aided and accelerated by computer-aided correlations. There is no need to consider yet again the detailed history of the development of the periodic table except to note, and this was the point of the above preamble, that progress was slowed by a persistent suspicion that the topic was tainted by mysticism; particularly that of Pythagoras[14]. After all, Newton himself had been inspired by the latter musical nonsense to claim that the rainbow consists of just 7 colors; a particularly silly idea that persists – and is *taught* – to this very day. Even in the 21st century, following a review of the mathematical structures such as number theory, information theory, order theory, set theory and topology unpinning the periodic law, a rather mawkish harking-back to the Pythagorean doctrine can still be found[15].

But the main problem, in believing that the elements bore a simple integer relationship to each other, was certainly that the measured Daltonian combining weights were affected by inadequate purification. Later confusion was further sown by the then-unknown existence of isotopes. Even with these drawbacks, ardent believers found meaningful patterns among the data; such as Dobereiner's triads[16], the 'telluric screw' of De Chancourtois[17] and Newland's octaves[18]. The latter must have been viewed with particular suspicion by those who were wary of Pythagorean musical mysticism. In spite of several rival elemental tables being proposed at the time[19,20], and afterwards[21], Mendeleev's table[22] now receives universal credit and currency largely because of his use of it to *predict* the existence of then-unknown elements. The continued use of modifications of the table to predict the existence of, and properties of, potentially useful new materials is described in what follows.

1 He 0.00	2 Ne 0.04	3 Ar 0.08	4 Kr 0.12	5 Xe 0.16	6 Rn 0.20	7 Fr 0.23	8 Cs 0.25	9 Rb 0.30	10 K 0.35
20 Sc 0.67	19 Y 0.66	18 Eu 0.655	17 Yb 0.645	16 Ca 0.60	15 Sr 0.55	14 Ba 0.50	13 Ra 0.48	12 Li 0.45	11 Na 0.40
21 Lu 0.675	22 Tm 0.678	23 Er 0.68	24 Ho 0.683	25 Dy 0.685	26 Tb 0.688	27 Gd 0.69	28 Sm 0.693	29 Pm 0.695	30 Nd 0.698
40 Bk 0.723	39 Cf 0.72	38 Es 0.718	37 Fm 0.715	36 Md 0.713	35 No 0.71	34 Lw 0.708	33 La 0.705	32 Ce 0.702	31 Pr 0.70
41 Cm 0.725	42 Am 0.728	43 Pu 0.73	44 Np 0.733	45 U 0.735	46 Pa 0.738	47 Th 0.74	48 Ac 0.743	49 Zr 0.76	50 Hf 0.775
60 Mn 0.945	59 Re 0.94	58 Te 0.935	57 Cr 0.89	56 W 0.885	55 Mo 0.88	54 V 0.84	53 Ta 0.83	52 Nb 0.82	51 Ti 0.79
61 Fe 0.99	62 Os 0.995	63 Ru 1.00	64 Co 1.04	65 Ir 1.05	66 Rh 1.06	67 Ni 1.09	68 Pt 1.105	69 Pd 1.12	70 Au 1.16
80 Al 1.66	79 In 1.60	78 Tl 1.56	77 Be 1.50	76 Zn 1.44	75 Cd 1.36	74 Hg 1.32	73 Mg 1.28	72 Cu 1.20	71 Ag 1.18
81 Ga 1.68	82 Pb 1.80	83 Sn 1.84	84 Ge 1.90	85 Si 1.94	86 B 2.00	87 Bi 2.04	88 Sb 2.08	89 As 2.16	90 P 2.18
100 N 3.00	99 Cl 2.70	98 Br 2.64	97 I 2.56	96 At 2.52	95 C 2.50	94 S 2.44	93 Se 2.40	92 Te 2.32	91 Po 2.28
101 O 3.50	102 F 4.00	103 H 5.00							

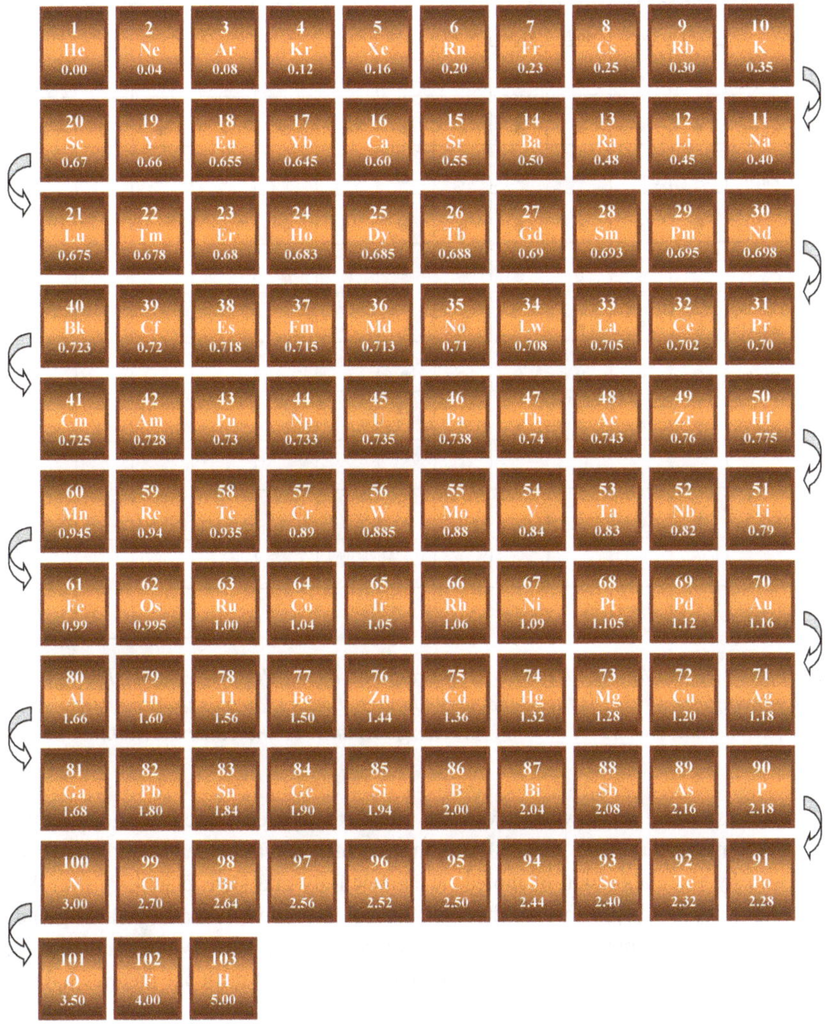

Figure 1. The Pettifor Scheme for ordering the elements with regard to compound structure-prediction. The upper number is termed the Mendeleev Number and the lower figure is the Chemical Scale value.

3

	Ti,51	Ta,53	Mo,55	W,56	Ni,67	Pt,68	Ag,71	Cu,72	Mg,73	Zn,76	Al,80	Pb,82
Pb,82	black	black	black	black	blue	black	black	black	black	black	black	blue
Al,80	black	black	black	black	black	blue	black	black	blue	blue	brown	
Zn,76	blue	blue	blue	blue	blue	blue	blue	blue	black	brown		
Mg,73	black	blue	black	black	blue	blue	blue	blue	brown			
Cu,72	blue	blue	black	blue	blue	brown	brown	red				
Ag,71	blue	blue	blue	blue	blue	brown	brown					
Pt,68	blue	blue	blue	blue	brown	blue						
Ni,67	blue	blue	blue	blue	brown							
W,56	blue	blue	blue	blue								
Mo,55	blue	blue	blue									
Ta,53	blue	brown										
Ti,51	brown											

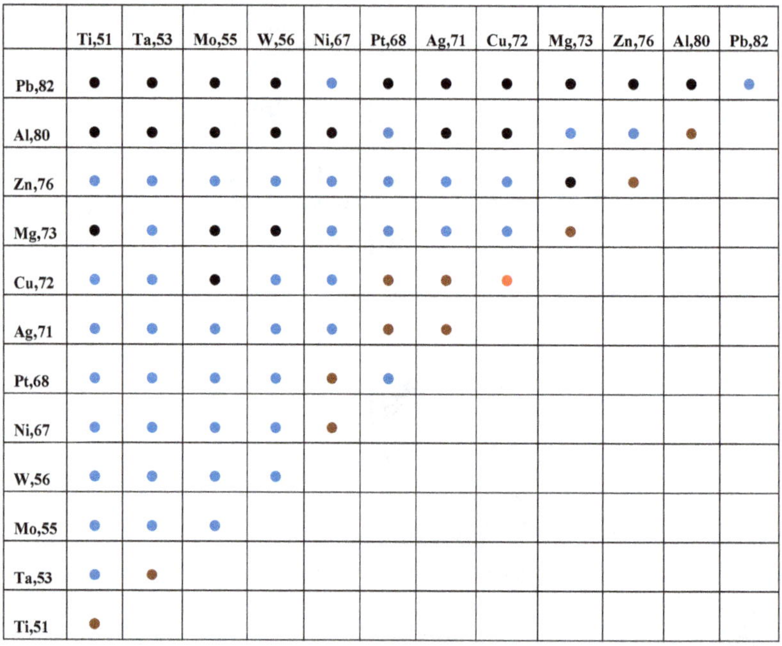

Figure 2. Weldability of elements as predicted by the Pettifor scheme. The figure following the symbol is the Mendeleev number. Black indicates unsatisfactory, blue indicates poor, brown indicates good and red indicates excellent.

The understanding of how the various elements formed structures grew with the introduction, starting in about 1850, of concepts such as valence and bonding. This was followed by the elaboration of molecular connectivity and stereochemistry even before the 1925 genesis of quantum mechanics. The valence bond method of Pauling and the molecular orbital methods of Bloch, Hückel, Hund and Mulliken aided later developments. Free electron, pseudopotential and density functional theories have been very successful in comprehending the structures and bonding of solids. In parallel with these concrete computational methods there have also existed rule-of-thumb predictive approaches such as those of Zintl and Hume–Rothery.

Materials Research Forum LLC
https://doi.org/10.21741/9781644902011

The Mendeleev Number and Structure-Mapping

In spite of the useful insights into chemical similarities and quantum-mechanical aspects which are provided by the familiar arrangement of elements in the essentially 2-dimensional Periodic Table, and in some less popular 3-dimensional arrangements, the mathematician David G.Pettifor took the seemingly outrageous step in the 1980s of 'unravelling' the classic Mendeleev tabulation so as to produce a 1-dimensional string.

	Ti	Ta	Mo	W	Ni	Pt	Ag	Cu	Mg
phosphor-bronze	•	•	•	•	•	•	•	•	•
bronze	•	•	●	•	•	•	•	•	
Cu-Be	•	•	•	•	•	•	•	•	
brass	•	•	●	●	•	•	•	•	
Ni-Ag	•	•	•	•	•	•			
Inconel	•	•	•	•	•				
Ni-Chrome	•	•	•	•					
mild steel	•	•	•	•					
stainless steel	•	•	•	•					
Manganin	•	•	•	●					

Figure 3. Weldability of elements to alloys as predicted by the Pettifor scheme. Black indicates unsatisfactory, blue indicates poor and brown indicates good.

The justification for this was that, although the table is invaluable in many respects, a simpler 1-dimensional ordering could place chemically similar elements in adjacent positions. This order has since proved to be very useful in the field of data-mining and accelerated materials design. The original stimulus was the prediction of the structural stability of AB compounds which, at the time, were known to exhibit one of 34 structures. By using A and B elemental types as coordinates, the various structures could be mapped out on a 2-dimensional plot. Pettifor aimed to order the elements so as to obtain the optimum separation of structures within the structure-map. Previous attempts to achieve this had, in effect, used the familiar Hume-Rothery parameters for alloying:

that is, the atomic radius, electronegativity and numbers of valence electrons. This necessitated the use of multi-dimensional plots and, more relevantly, yielded poor structural separation. Such two-dimensional structure-maps had been restricted to particular classes, such as *sp*-bonded octets. In order to demarcate structures having a given stoichiometry, it had been necessary[23] to plot 3-dimensional arrangements using coordinates of electronegativity, atom size and number of valence electrons; sometimes leading to 16 layers of 2-dimensional plots. Problems naturally arose because the electronegativity scale assigned similar values to diverse elements such as zirconium, cobalt and gallium while the atom sizes varied less across groups than down groups. As for the average number of valence electrons, this criterion did not distinguish transition metals from p-bonded elements to their right; essentially because of a neglect of detailed quantum mechanical differences.

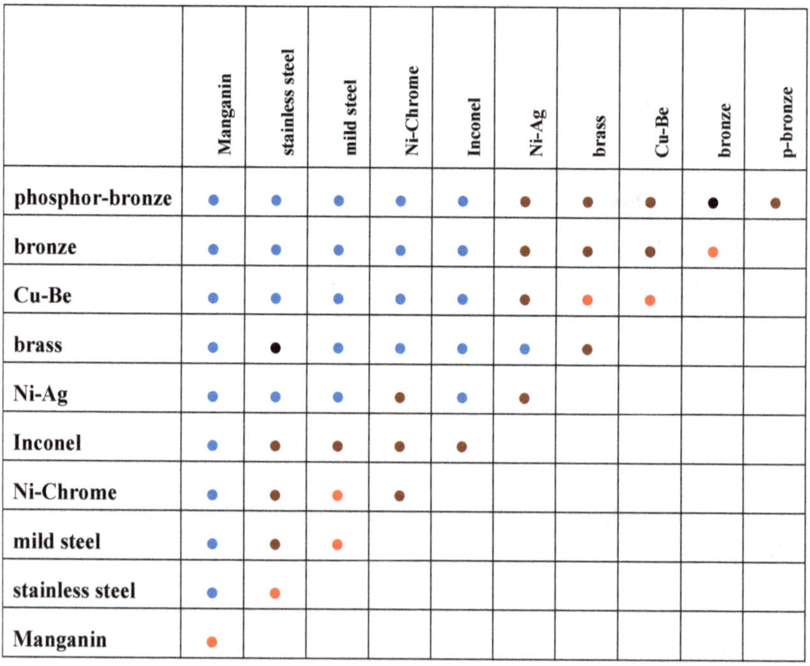

	Manganin	stainless steel	mild steel	Ni-Chrome	Inconel	Ni-Ag	brass	Cu-Be	bronze	p-bronze
phosphor-bronze	●	●	●	●	●	●	●	●	●	●
bronze	●	●	●	●	●	●	●	●	●	
Cu-Be	●	●	●	●	●	●	●	●		
brass	●	●	●	●	●	●	●			
Ni-Ag	●	●	●	●	●	●				
Inconel	●	●	●	●	●					
Ni-Chrome	●	●	●	●						
mild steel	●	●	●							
stainless steel	●	●								
Manganin	●									

Figure 4. Weldability of alloys as predicted by the Pettifor scheme. Black indicates unsatisfactory, blue indicates poor, brown indicates good and red indicates excellent.

Pettifor's solution was quite draconian in that he dismissed all of the existing theoretical precepts and instead chose a purely phenomenological ordering of the elements that ensured an almost perfect demarcation of stoichiometric AB structures. The Pettifor Scale is shown in figure 1. The so-called chemical scale, χ, ordered the elements along a single axis and the latter were assigned a Mendeleev number. The chemical scale was completely phenomenological and served only to order the elements relative to one another. Note nevertheless that the scale also includes very exotic entries which are unlikely to be relevant to the development of commercial products. It was reassuring that it was later found that the

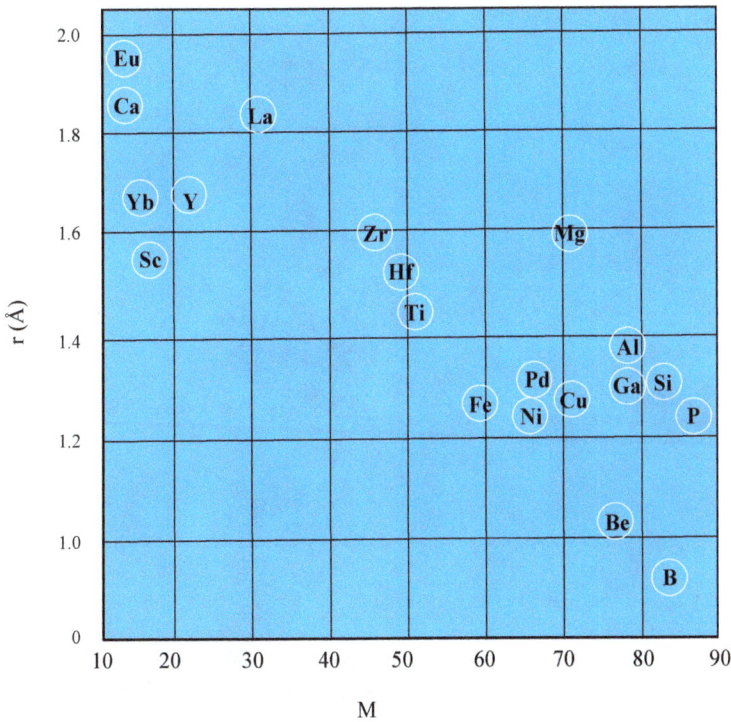

Figure 5. Radius of atoms versus the Mendeleev number of elements which are relevant to the formation of quasicrystals and bulk metallic glasses.

original *ad hoc* ordering also neatly demarcated the structures of non-stoichiometric AB systems[24], and maps were plotted for AB_2, AB_3, AB_4, AB_5, AB_6, AB_{11}, AB_{12}, AB_{13}, A_2B_3,

A_2B_5, A_2B_{17}, A_3B_4, A_3B_5, A_3B_7, A_4B_5, and A_6B_{23}; even FeB (B27) and CrB (B33) could be demarcated[25,26]. It was further shown[27] that it was possible to order binary and ternary alloys, with A_2B_{17}, AB_{11}, AB_{12} and AB_{13} stoichiometries, within a single 2-dimensional structure-map.

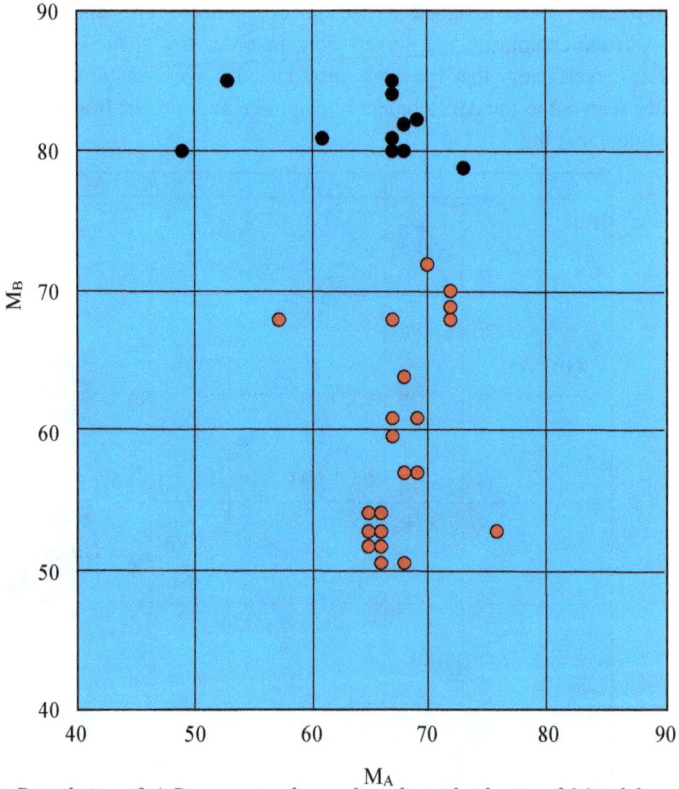

Figure 6. Ductilities of A_3B compounds predicted on the basis of Mendeleev numbers. Black indicates brittleness and red indicates ductility. The compounds are Ni_3Al, Pt_3Ti, Cu_3Au, Mg_3In, Ni_3Si, Pt_3Pb, Zr_3Al, Pd_3Fe, Ni_3Ga, Pd_3Pb, Au_3Cu, Ni_3Pt, Fe_3Ga, Rh_3Ta, Zn_3Nb, Nb_3Si, Ni_3Ge, Ir_3V, Pt_3Co, Rh_3Nb, Cr_3Pt, Rh_3V, Pd_3Cr, Ir_3Nb, Ir_3Ta, Ni_3Fe, Pt_3Cr, Pt_3Al, Ir_3Ti, Ni_3Mn, Cu_3Pt and Cu_3Pd. If necessary, an individual compound can be identified by noting that discernible rows going down the diagram correspond to Si, Ge, Pb, Ga, Al, In, Cu, Au, Pd, Pt, Co, Fe, Mn, Cr, V, Nb, Ta and Ti in order while discernible columns going across the diagram correspond to Zr, Nb, Cr, Fe, Rh, Ir, Ni, Pt, Pd, Au, Cu and Zn in order.

The Pettifor chemical scale also helped to address more chemically related questions. The stability of the local fourfold coordination of divalent and trivalent metal ions in liquid mixtures of polyvalent metal halides and alkali halides was generally classified in terms of structural coordinates which were deduced from properties of the elements such as nodal or pseudopotential radii. As an alternative, a classification which was based upon Pettifor's chemical scale was considered[28]. The classifications which were based upon elemental properties were generally successful in distinguishing those molten mixtures in which experimental evidence indicated a long-lived fourfold coordination of polyvalent metal ions, while Pettifor's scale was useful for studying the finer details of local

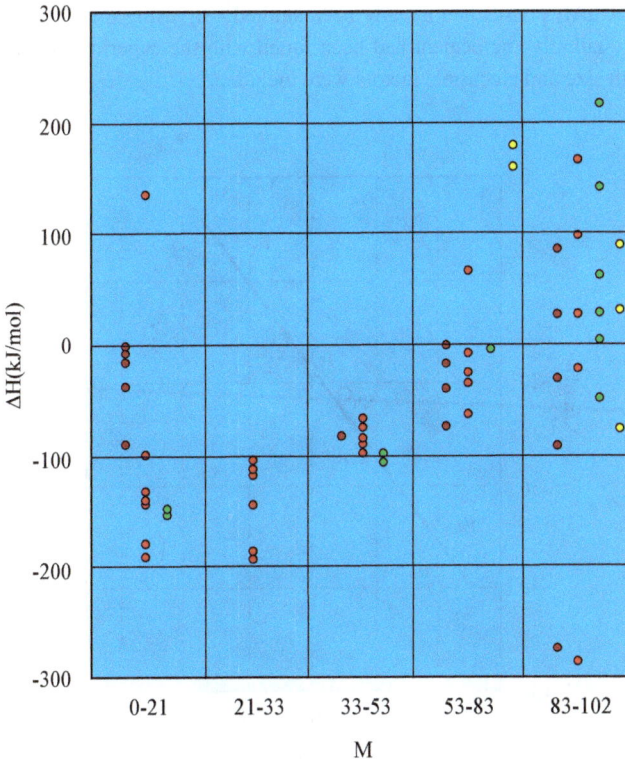

Figure 7. Enthalpy of formation of metal hydrides, MH_x, versus the Mendeleev number. Brown: $x = 1$, red: $x = 2$, green: $x = 3$, yellow: $x = 4$.

coordination in the liquid state. In addition, divalent and trivalent metal ions in stoichiometric liquid mixtures of their halides were often four-fold or six-fold coordinated by halogens into quite long-lived complexes. The local coordination in about 140 mixtures was successfully classified[29,30] by using a structure-sorting method which was based upon the Pettifor chemical scale.

The validity of Pettifor predictions for Laves phases in pseudobinary systems with transition elements was confirmed[31] by experimental results for the Zr-Nb-Fe system. In addition to the C15-type $(Zr,Nb)Fe_2$ phase, a C14-type Laves phase, $Zr(Nb,Fe)_2$, was found. The differing roles played by atomic size and electronic factors in stabilizing transition-metal Laves phases with respect to the competing phases of C11b-type $MoSi_2$ and C16-type $CuAl_2$ phases had already been studied[32] by using a tight-binding d-bond model. Good qualitative agreement had been found with the experimental AB_2 structure map when both size and electronic factors were included.

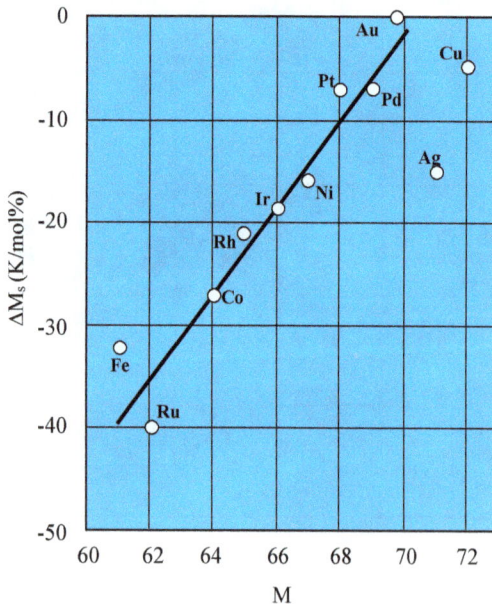

Figure 8. Change in the martensite start temperature of Ti-Au alloys when 1mol% of a ternary element with a given Mendeleev number is added.

Materials Research Forum LLC
https://doi.org/10.21741/9781644902011

When Pettifor structure mapping was applied[33] to interstitially stabilized AB_3C-Cu_3Au intermetallics, it was incidentally noted that binary alloys which exhibit complete solid solubility fall into the range of Mendeleev numbers between 50 and 70.

In more recent times, new equi-atomic ABX compounds were predicted[34] to exist, where A and B were metals and X was silicon or germanium, by looking for regularities in databases. Computer analysis showed that the factors which were most important for predicting the formation, or not, of equi-atomic compounds were the Mendeleev numbers of element A or B, and the thermal conductivities of element A or X.

The original Pettifor structure-maps were soon exploited[35] by alloy-designers for the prediction of new ternary or quaternary systems having a structure, such as Cu_3Au, which would impart good high-temperature mechanical properties or to predict the weldability of different metals[36] (figure 2) and alloys (figures 3 and 4). Later work extended the predictive power of the maps to more complicated multi-component systems[37]. It was noted[38] that the Pettifor approach indicated that many higher-order alloys can be thought of as being pseudo lower-order alloys. It was demonstrated that most quasicrystals are pseudo-binary and that most metallic glasses are pseudo-ternary. By considering[39] the largest atom to be the most important constituent of a material, it was shown that most ternary and quaternary quasicrystals could be treated as pseudo-binary intermetallics. This led to a grouping of quasicrystals into 4 structural classes on the basis of the bond orbital of the large atom (figure 5). The new classification of quasicrystals was centered on lithium, magnesium, aluminium, gallium, calcium, scandium, yttrium, titanium, zirconium, hafnium and rare earth elements, given that they were the largest atoms.

It was recalled that the Pauling electronegativity had been suggested to be an indicator of the mechanical behavior of polycrystalline $L1_2$ intermetallic compounds, but it was shown[40] that a map based upon the electronegativities of the components was unable to discriminate between ductile and brittle polycrystalline materials. For example, intergranularly-brittle Ni_3Al and ductile Ni_3Mn necessarily mapped to the same spot. The Pauling scale, and any scale that did not distinguish directional from isotropic electronegativities, was thus unsuitable. The structure map was instead based upon the then-novel Mendeleev number of Pettifor. It was found that the $L1_2$ compounds then separated into 2 distinct groups (figure 6). The Mendeleev number was seen to encompass several aspects of chemical activity in addition to the electronegativity. One quibble was that the Mendeleev number was fixed while the effective s-orbital electronegativity could vary; offering an additional degree of freedom. That is, a material like monocrystalline Ni_3Al might be ductile while polycrystalline Ni_3Al could be quite brittle.

Algorithms have been developed for turning the data from a crystal-structure database into a Pettifor map[41]. These were based upon a list of candidate crystal structures. In the case of AB and A_3B compounds it was found that, for a new unknown alloy having a stable structure at the stoichiometry of the Pettifor map, a candidate list of just 5 structures included the correct one in 86% of cases.

Pettifor structure-mapping has been applied[42] to MH_x binary metal hydrides with regard to defining their hydrogenation properties. The effect of the formation enthalpy upon the map was studied by plotting the enthalpy versus the Mendeleev number and the chemical scale. Higher values of the Mendeleev number indicated a lower hydride stability. Metal hydrides containing only one hydrogen atom, and having a Mendeleev number of between 0 and 21 exhibited adequate stability (figure 7). When the Mendeleev number was between 21 and 53, all of the hydrides exhibited a high stability regardless of x. When the Mendeleev number was between 53 and 102, the hydrides exhibited large variations in stability. In general, hydrides with x = 1 to 3 exhibited high or adequate stability, and those with x = 4 were highly unstable.

Table 1. Atomic radius mismatch, binary heat-of-mixing and
Mendeleev Number of components of glass-forming alloys

Component 1	Component 2	Mismatch (%)	Heat-of-Mixing (kJ/mol)
Ti (51)	Cu (72)	13	-9
Ti (51)	Ni (67)	15	-35
Zr (49)	Cu (72)	20	-23
Zr (49)	Ni (67)	22	-49
Hf (50)	Cu (72)	19	-17
Hf (50)	Ni (67)	21	-42
Ti (51)	Zr (49)	9	0
Ti (51)	Hf (50)	7	0
Zr (49)	Hf (50)	1	0
Cu (72)	Ni (67)	2.5	4

H 0.8 6303	O 0.8 18805	Xe 1.0 98	F 2.1 3072	C 3.0 4360	N 3.1 4366	B 5.4 2278	P 9.6 4165	Li 10.1 1947
Na 10.8 2729	Cl 11.0 2902	Be 11.5 416	S 11.6 4934	I 12.3 1409	Cd 13.7 1067	Ba 16.2 2383	K 16.4 3019	Mo 16.9 1257
Cu 18.0 2806	Re 18.2 457	Cs 19.0 1916	Te 19.1 1582	Si 19.7 3743	Hg 20.3 680	Bi 21.3 1143	V 22.7 1442	W 24.3 683
Sb 24.6 1722	Al 24.7 2202	Ag 24.8 1103	Kr 25.0 8	Se 26.6 2212	Ca 27.2 1800	Pb 27.7 1115	As 28.3 1634	Zn 29.5 1513
U 29.6 1028	Nb 29.8 998	Sn 30.3 1691	Mg 31.0 1289	Br 31.0 1136	In 31.0 1440	Ti 32.8 1203	Tl 33.3 847	Au 34.2 825
Sr 34.3 1569	Ge 35.2 2149	Ta 35.4 687	Tc 35.5 65	Fe 37.0 2285	Mn 37.1 1929	Rb 37.1 1404	Cr 37.1 1054	Ga 38.1 1269
Ni 40.0 2188	Ru 42.3 770	Os 43.0 328	Co 44.0 1888	Pd 44.3 1135	Zr 44.6 865	La 46.5 6303	Pt 47.4 900	Sc 47.9 597
Es 50.0 2	Ir 52.3 589	Th 54.2 406	Rh 55.1 813	Ce 58.1 1005	Eu 58.3 676	Y 59.9 936	Yb 65.7 682	Hf 66.9 405
Np 67.9 168	Nd 70.7 933	Pu 71.2 184	Cm 73.5 49	Pr 74.5 773	Lu 75.3 417	Gd 75.9 784	Er 76.1 686	Pa 77.8 27
Sm 78.6 678	Bk 78.6 14	Po 78.6 28	Ho 83.5 613	Tm 83.8 407	Tb 83.8 647	Dy 86.3 606	Cf 92.9 14	Am 93.2 44
Ac 100 5	Pm 100 6	Ra 100 3						

Figure 9. Using the Inorganic Crystal Structure Database, the first figure indicates the percentage replaceability of the element. That is, it represents the probability that a new stable crystal structure is obtained by replacing that element. The second number is the total known number of relevant compounds in the database.

The predictions were not always successful. In an early attempt to improve the ductility of Ti_3Al by alloying, in order to change the crystal structure from ordered hexagonal, $D0_{19}$, to face-centered cubic $L1_2$, scandium was partially substituted[43]. If it was assumed that all the scandium atoms were located in the titanium sub-lattice, an average Mendeleev number for the titanium sub-lattice of scandium-added alloys could be calculated. The average Mendeleev number of the titanium sub-lattice was less than 49. According to Pettifor-map predictions, the modified alloys should have exhibited an $L1_2$ structure. They instead exhibited $D0_{19}$ structures.

The Mendeleev number does help to clarify the glass-forming tendency of ternary copper-containing titanium, zirconium and hafnium alloys. When those having the compositions $Zr_{25}Ti_{25}Cu_{50}$, $Zr_{34}Ti_{16}Cu_{50}$, $Zr_{25}Hf_{25}Cu_{50}$ and $Ti_{25}Hf_{25}Cu_{50}$ were rapidly solidified[44] so as to produce ribbons, all of them were amorphous. In the case of $Zr_{34}Ti_{16}Cu_{50}$, localized precipitation of Cu_5Zr occurred and the alloys tended towards phase separation in the initial stages of crystallization. The difference in the crystallization behaviors of the alloys, as compared with that of nickel-bearing ternary alloys could be explained in terms of atomic size, binary heat of mixing and Mendeleev number (table 1). The atomic-radius mismatches, with copper, of titanium, zirconium or hafnium are 13, 20 and 19%, respectively, and the corresponding atomic-radius mismatches, with zirconium, of titanium and hafnium are 9 and 1%, respectively. The heats of mixing, with nickel, of titanium, zirconium and hafnium are very negative. As the atomic size difference between zirconium and hafnium is small, when compared with that between zirconium and titanium, the replacement of titanium by hafnium does not affect the glass-forming ability of the alloy. The heat-of-mixing of titanium, zirconium or hafnium with nickel is more negative than that with copper. So although copper and nickel appear next to one another in the periodic table, their thermodynamic behaviors with respect to these other elements are different. In this regard again, the Mendeleev number can be used to rationalize the differences: their atomic numbers are next to one another but their Mendeleev numbers are not, thanks to Pettifor's 'unravelling' of the periodic table.

Original	He	Ne	Ar	Kr	Xe	Rn	Fr	Cs	Rb	K	Na	Li	Ra
Modified	He	Ne	Ar	Kr	Xe	Rn	Fr	Cs	Rb	K	Na	Li	Ra

Ba	Sr	Ca	Yb	Eu	Y	Sc	Lu	Tm	Er	Gd	Sm	Pm	Nd
Ba	Sr	Ca	Eu	Yb	Lu	Tm	Y	Er	Ho	Sm	Pm	Nd	Pr

Pr	Ce	La	Lr	No	Md	Fm	Es	Cf	Bk	Cm	Am	Pu	Np
Ce	La	Ac	Th	Pa	U	Np	Pu	Am	Cm	Bk	Cf	Es	Fm

U	Pa	Th	Ac	Zr	Ta	V	Mo	W	Cr	Tc	Re	Mn	Fe	Os
Md	No	Lr	Sc	Zr	Nb	V	Cr	Mo	W	Re	Tc	Os	Ru	Ir

Ru	Co	Ir	Rh	Ni	Pt	Pd	Au	Ag	Cu	Mg	Hg	Cd	In	Al
Rh	Pt	Pd	Au	Ag	Cu	Ni	Co	Fe	Mn	Mg	Zn	Cd	Ga	In

Ga	Pb	Sn	Ge	Si	B	Bi	Sb	As	P	Po	Te	Se	S	C	At
Tl	Pb	Sn	Ge	Si	B	C	N	P	As	Sb	Bi	Po	Te	Se	S

I	Br	Cl	N	O
O	At	I	Br	Cl

*Figure 10. Comparison of the original and modified Pettifor scale,
with highlighted differences*

The Mendeleev number also rationalized[45] the effect of ternary additions upon the martensitic transformation start temperature, M_s, of TiAu alloys. These ternary elements, iron, cobalt, nickel, copper, ruthenium, rhodium, palladium, silver, iridium and platinum all replaced the gold to the extent of between 1 and 3mol%. The room-temperature phase was B19, regardless of the nature of the ternary additions. Correlations were found between M_s, or its modification, and the electron/atom ratio, valence-electron number, Mendeleev number (figure 8) and c/b lattice-parameter ratio of the B19 martensite phase. The M_s temperature decreased linearly with increasing amount of ternary addition. A similar study had been made[46] of the effect of ternary elements upon the martensite start temperature of TiNi shape-memory alloys. The ternary elements were taken from the 4th-period group (Zr, Hf) to the 10th-period group (Pd, Pt), and the changes were correlated with the number of total outer d- and s-electrons, the electron hole number, the electronegativity, the atomic volume and the Mendeleev Number. The change in start temperature depended upon the number of total outer d- and s-electrons and the electron hole number, but the data were scattered and other factors were clearly involved. The change in M_s appeared to decrease slightly with increasing electronegativity, and also depended weakly upon the atomic volume. Ternary additions having a large atomic volume slightly decreased the change in start temperature slightly, and *vice versa*. The change in M_s was closely related to the Mendeleev number of the addition, with a slope of -9.4K/mol%; suggesting that a ternary element having a smaller number stabilized the B19' martensite while one having a larger number stabilized the B2 parent phase.

At the time when the Pettifor scheme was developed, the available data points were relatively limited in number[47]. The modern, hundred-fold larger database, permits more detailed correlations to be found. The Pettifor concept, that chemical similarity leads to the formation of similar structures, has been further explored[48]. One check is to determine how often the replacement of an element in a given compound produces another compound having the same type of structure.

This led to the estimation of the replaceability of a given element (figure 9). The highest replaceabilities are found for lanthanides and actinides, while the lowest values are found for hydrogen and other first-row elements. The significance of the replaceability results should obviously be treated with some caution when the sample space is very limited. This work also led to the development of a modified Pettifor scheme (figure 10).

Hundreds of novel AB_2X_2 compounds, where A and B are various elements and X is boron, aluminium, silicon, phosphorus, gallium, germanium, arsenic, tin or antimony, have been computer-designed and their crystal structures predicted. This was done by using a set of databases on the properties of inorganic substances, together with pattern-recognition techniques. It is found[49] that the possible formation of AD_2X_2 compounds and their structures depend largely upon the Mendeleev numbers and pseudopotential radii. With regard to the miscibility of binary alloys, the mining of data on hundreds of experimental phase diagrams and thousands of thermodynamic data-bases, yielded[50] a classification of alloying behavior for 813 binary alloys involving transition and lanthanide metals. It was again noted that a slightly-modified Pettifor chemical scale provided a unique 2-dimensional map which demarcated miscible and immiscible systems. The miscibility map also strongly confirmed the value of the Miedema theory, which has sometimes been quite strongly criticized[51].

The Descriptor Problem

When mining large databases for correlations between properties and structure it is important to identify suitable parameters, *descriptors*, which succinctly characterize the properties of a material as this aids computer-processing. It has been pointed out[52] however that, when the scientific basis for the relationship between the descriptor and the precise mechanism involved is unclear, the reason for the descriptor-property link is uncertain. Superior criteria were therefore defined for choosing a suitable descriptor. It was demonstrated how a meaningful descriptor could be found systematically by using, as an example, the energy difference between zincblende or wurtzite and rock-salt semiconductors. As a proof-of-concept, it was shown[53] elsewhere how to construct a physical model for the quantitative prediction of the crystal structure of binary compound semiconductors.

A systematic approach was proposed[54] for discovering descriptors within the framework of compressed-sensing based dimensionality-reduction. This so-called *sure independence screening and sparsifying operator* could handle huge and correlated parameter-spaces, and could converge to an optimal solution after starting from a combination of features which were relevant to materials properties that were of interest. The method also yielded stable results while using small training sets. The method was tested by making quantitative predictions of the ground-state enthalpies of octet binary materials while making use of *ab initio* data. It was applied to the prediction of the metal/insulator classification of binary compounds. Accurate predictive models were found in each case. In the case of the metal/insulator classification model, the predictive capability was tested to beyond the training data used. This reproduced available pressure-induced insulator-to-metal transitions and permitted the prediction of as-yet unknown candidates for experimental testing.

Clustering Methods

A method was presented[55] for calculating the phase diagrams of isostructural solid alloys from a microscopic theory of electronic interaction. The internal energy of the alloy was first expanded as a series of volume-dependent multi-atom interaction energies which were deduced from self-consistent total-energy calculations for periodic compounds, performed within the local-density formalism. Distant-neighbor interactions were then renormalized into composition-dependent and volume-dependent effective near-neighbor multisite interactions. Approximate solutions to the general Ising model were finally obtained by using the tetrahedron cluster variation method. These generated the excess enthalpy, entropy and … finally the phase diagram. The method was applied to prototype semiconducting face-centered cubic alloys. One of these, $GaAs_xSb_{1-x}$, had a large size-mismatch while a second, $Al_{1-x}Ga_xAs$, had a small size-mismatch. The predictions were in excellent agreement with measured miscibility-temperature and excess-enthalpy data. In lattice-mismatched systems, it was found that O < HO < HD, where O was an ordered Landau-Lifshitz structure and D was a disordered phase. It was thus predicted that such an alloy would disproportionate at low temperatures into the binary constituents. If instead disproportionation was kinetically inhibited, an ordered phase such as chalcopyrite would be thermodynamically more stable, at below a critical temperature, than was the disordered phase having the same composition. In lattice-matched systems, it was found that O < HD < HO for all Landau-Lifshitz structures so that phase-separating behavior alone was predicted. In those systems, longer-period ordered superlattices were more stable, at low temperatures, than was the disordered alloy.

Solid-solid and solid-liquid phase equilibria in the Ni-Ti system were studied[56] within the framework of the cluster variation method. The energy parameters involved in the free-energy description of each phase were determined by performing tight-binding energy calculations using the cluster Bethe lattice method. The pair interactions were restricted to first-nearest neighbors in face-centered cubic based structures and to first- and second-nearest neighbors in body-centered cubic based structures. The configurational entropies for disordered solutions and for ordered compounds were obtained by using the tetrahedron approximation in the cluster variation method. Three compounds, NiTi, Ni_3Ti and $NiTi_2$, were treated as stoichiometric compounds and the calculated diagram agreed reasonably well with that found experimentally. In further work[57], tight-binding electronic band structure calculations were combined with a free-energy expression arising from the cluster variation method in order to study phase equilibria in the Ni-Ti system. The effective pair interactions which were used in the cluster variation calculations were evaluated by using the generalized perturbation method. The configurational entropy for disordered solutions and for ordered compounds were again obtained by using the tetrahedron approximation of the cluster variation method. The predicted phase diagram was again in good agreement with the experimentally determined one.

The phase stability of superstructures, in the Au-Pd and Ag-Pt alloy systems, which were based upon the face-centered cubic lattice were studied[58] from a fully relativistic electronic density functional theory point of view. Electron-ion interaction was described by using the projector augmented-wave method, and exchange-correlation effects were treated in the generalized gradient approximation. The cluster expansion method was used to obtain effective cluster interactions in the face-centered cubic lattice and guided a systematic ground-state search of both alloy systems. The ground-state analysis revealed many ground states in Au-Pd; especially on the gold-rich side. Possible long-period superstructures occurred near to the $Au_{70}Pd_{30}$ composition. Ground-state analysis further indicated the existence of a particularly stable AgPt compound having the $L1_1$ (CuPt) structure. It also suggested the existence of a marginally stable ordered compound, Ag_3Pt. The study ruled out however the existence of the very stable Ag_3Pt phase, with the $L1_2$ structure, which had long been accepted. No sign was found of a stable ordered state at the $AgPt_3$ composition. The cluster variation method, with a large maximum cluster, was used to calculate the enthalpy-of-mixing of the disordered solid solutions and of the solid portions of the Au-Pd and Ag-Pt phase diagrams. It was concluded that cluster expansions could not account for the high-temperature miscibility gap in the Ag-Pt system if the effective cluster interactions did not extend beyond second-nearest neighbor. Only when third-nearest neighbors were included in the cluster expansion did it become possible to obtain a phase diagram that agreed qualitatively with the known Ag-Pt phase diagram. The 0K enthalpies of metallic elements having several tetrahedral close-packed structures were calculated[59] using electronic density functional theory. A primitive version of a generalized cluster expansion was used to determine the feasibility of predicting ground-state structures within the class of tetrahedral close-packed structures and assessing consistency within sets of enthalpy data.

A crystal-structure predictor, based upon an evolutionary algorithm, generated hundreds to thousands of possible structures[60]. In order to aid the selection of possibly interesting structures, routine high-dimensional classification concepts were applied. Experiments were performed using various crystal-structure descriptors, distance measures and clustering methods. A visual design and validation method was used to build a library of structures and this helped to accelerate the analysis of the above crystal-structure predictor output by at least an order of magnitude.

The prediction, using first-principles calculations, of whether mixtures of metallic elements phase-separate or form ordered structures is a problem which can be addressed, using cluster-expansion and various search algorithms, when experiment suggests that the underlying lattice is conserved. Evolutionary algorithms have been used[61] to search for stable off-lattice structures at fixed compositions. An integrated approach, involving cluster-expansion and high-throughput first-principles calculations, have been applied to the full range of compositions in binary systems where the end-members or intermediate ordered compounds had differing lattice types. This approach replaced search algorithms that involved the direct calculation of a modest number of naturally occurring prototypes which represented all crystal systems and informed cluster-expansion calculations of structures. The approach maintained the precision of the cluster-expansion method and the strengths of the first-principles method. Its application to poorly-understood binary hafnium systems which were believed to be phase-separating, indicated 3 classes of alloy where cluster-expansion and first-principles complemented one another and could reveal new ordered structures.

In addition to the innovative Mendeleev number concept, the Pettifor group has followed parallel lines of enquiry. The relative stabilities of the 7 commonest structures of pd-bonded AB compounds were studied[62] by using a tight-binding model which accounted for the effect of atomic size, of the separation of the p and d energy-level and of band-filling[63]. This theory, although ignoring valence s-electrons, could qualitatively demarcate the domains of NaCl, CsCl, NiAs, MnP and boride stability; leaving just the narrow FeSi domain unexplained. An analytical many-body potential was developed[64], for the bond-order of s-valent systems, which directly accounted for the effect of the local atomic environment upon the bond strength. The influence of 3-membered- and 4-membered ring terms upon the topology of 4-atom clusters was demonstrated, and the simplicity of the methods used permitted embedded-atom potentials to be extended so as to include 3-body and 4-body terms. The relative stability of 3-, 4-, 5- and 6-atom s-valent clusters was investigated[65] by using a nearest-neighbor tight-binding Hückel model. The predicted trends in structure type as a function of electron count were directly related to the cluster topology by using a ring approximation to the bond order, and it was noted that cluster connectivity was more important a factor than was symmetry. A simple nearest-neighbor orthogonal tight-binding model could also qualitatively explain[66] the trends in structural stability among sp-bonded elements. Relative stabilities were predicted by directly comparing the band energies by using the structural energy difference theorem and the bond lengths. This then permitted the structural trends to be interpreted in terms of the topology of the local atomic environment as reflected by the behavior of the first few moments of the density-of-states.

The relative structural stability of s- and sp-valent systems was again examined[67] by using a fourth-moment approximation to tight-binding bond-order potentials[68], preferably subject to a sum-rule constraint. Competition between the graphite, diamond and simple-cubic sp-valent lattices was modeled particularly well by the angularly-dependent bond-order potentials. Dependable atomistic simulation of defects in intermetallics indeed required the development of angularly-dependent interatomic potentials[69]. That is, it was predicted that the nearest competing phases to the $L1_0$ ground state of TiAl and the B2

Materials Research Forum LLC
https://doi.org/10.21741/9781644902011

ground state of NiAl were CrB (B33) and FeSi (B20) respectively; the structural stability of which was governed by directional pd bonding. A simple d-p tight-binding model was proposed[70], for treating transition-metal monocarbides, which could predict the observed crystal structures and – moreover – the (100) surface (1 x 1) relaxations. The model showed that a reversal in the surface relaxations between TiC(100) and TaC(100) was linked to the d-p band-filling and required the occurrence of d-d interactions between second-nearest neighbor metal atoms. Phase transitions in titanium have recently been analyzed[71] by using a bond-order potential. It was noted that atomic-scale studies of titanium-based alloys, conducted using methods such as density functional theory, had been limited to groups which contained only a few hundred atoms. In the new approach, the tight-binding approximation was coarse-grained to the electronic structure and a pertinent bond-order potential was developed by fitting it to the energies and forces which are involved in elementary deformations of such simple structures. The new potential predicts the structural properties of stable and defective titanium phases with an accuracy which is comparable to that of previous tight-bonding methods.

Interatomic bond-order potentials within a reduced tight-binding framework were generalized[72] to the case of multi-component *sp*-valent systems, so as rigorously to obtain expressions for the σ and π bond orders in chemically heterogeneous situations. The method was then applied[73] to a study of the factors which control the relative structural stability of s-valent 4-atom clusters with respect to packing as linear chains, squares, rhombi or tetrahedra. It was noted that interatomic potentials which were based upon the second-moment approximation to the local density of states or bond-order were unable to predict the most stable structure. That required information concerning the higher moments, which bond-order potentials naturally accounted for. Simplified expressions were then deduced for the σ and π bond-orders of *sp*-elements having half-filled valence shells. These reproduced the relative stabilities of open and close-packed silicon phases, and also clarified the concept of single, double, triple and resonant bonds in carbon systems. An analytical interatomic bond-order potential which depended directly upon the group number of an *sp*-valent element was derived[74] by extrapolating previous bond-order potentials, for group-IV elements, from half-filled occupancy by using a simple function for the upper bound on bond order. This interatomic potential predicted the structural trends, across *sp*-valent elements, which were found by means of tight-binding calculations and were observed experimentally. The new bond-order potentials naturally reflected the valence-dependent nature of the bonding. Atomistic modeling of materials by means of effective potentials required a reliable rule for the breaking and forming of interatomic bonds in various atomic environments. Bond-order potentials[75] provided just such a description of atomic bonding while also being computationally efficient when performing large-scale atomistic simulations.

As an example of practical application a study was made[76], using density functional theory calculations and pseudopotentials within the generalized-gradient approximation, of the lattice stability of the intermediate phases of the Sr-Si system. Nine compositions and 26 possible crystal lattices were considered. The calculated heats of formation of the various polymorphs were in excellent agreement with experimental data. The Sr_2Si, Sr_5Si_3 and SrSi phases were predicted to exhibit high-pressure transitions at 5.5, 19.9, 11.8 and 60GPa, respectively. The band-gap of the semiconducting Sr_2Si ground state structure was calculated to be 0.29eV. The bonding of the Sr-Si phases was mainly ionic, but there was some evidence for the formation of directional covalent bonds between neighboring silicon atoms in the most silicon-rich phases.

It was also asked[77] whether the topology of the energy-landscapes of alloy systems obeys any rules, and this idea was explored by examining a reduced energy surface and its density of states, using a one-dimensional characterizing parameter in the form of a vector in abstract multidimensional space, by defining a metric for the distance between two structures and by incorporating aspects of order and entropy. This approach could rationalize large crystal-structure databases and tailor algorithms for structure-prediction. The study confirmed that low-energy minima were clustered into compact regions of configuration space which were very limited in number. It was quantitatively proved that crystals try to adopt the simplest structure possible.

A problem of great practical interest is the occurrence of topologically close-packed phases in Ni-based superalloys. Previous methods had used the average number of holes or the centre-of-gravity of elemental d-bands to predict whether a given alloy would suffer from topologically close-packed phase formation. The latter 1-dimensional methods failed however when applied to complex new alloys, so a 2-dimensional plot was instead proposed[78] which mapped the average electron concentration versus a composition-dependent size-factor difference. This map could demarcate experimental data on topologically close-packed phases, in binary A-B transition-metal alloys, into well-defined but occasionally overlapping regions. The latter corresponded to structures such as A15, σ, δ, μ and Laves phases. Topologically close-packed phases, regardless of the number of constituents, were found to be located in the same regions of the structure map which harbored binary compounds of the same structural type. Density functional theory, tight-binding and analytical bond-order potential methods were subsequently combined[79] in order to investigate the structural trends among topologically close-packed phases. Density functional theory was used to calculate structural energy differences across the 4d and 5d transition-element series and the heats of formation of Mo-Re, Mo-

Materials Research Forum LLC
https://doi.org/10.21741/9781644902011

Ru, Nb-Re and Nb-Ru alloys. It was shown that the valence-electron concentration stabilized A15, σ and χ phases but destabilized μ and Laves phases. A 1-parameter d-band tight-binding model, together with the structural energy difference theorem, reproduced the observed density functional theory structural trends. The structural energy difference theorem also explained the effect of relative size differences upon the stability of the μ and Laves phases in binary systems. Analytical bond-order potential theory, using tight-binding integrals, converged towards the tight-binding structural energy difference curves as the number of moments in the bond-order potential expansion was increased. It was thus found that the fourth-moment contribution demarcated A15, σ and χ phases from μ and Laves phases; as already noted using empirical structure-mapping. This was attributed to the difference in the bimodality of the corresponding density-of-states that was caused mainly by distortions, in their coordination polyhedra, away from the ideal Frank-Kasper polyhedra. At least 6 moments were required in order to predict the structural trend: A15 $\Rightarrow$ σ $\Rightarrow$ χ. The contributions made by atomic size and electronic features to the structural stability of topologically close-packed phases, and transition-metal carbides, were further investigated[80]. The importance of electronic factors with regard to topologically close-packed phases was indicated by the density functional theory predictions that the A15, σ and X phases were stabilized between groups VI and VII of the transition metals, while the μ and Laves phases were destabilized. This dichotomy was again attributed to the bimodal shape of the electronic density-of-states. The importance of the size factor in topologically close-packed phases was illustrated by the heats of formation of Mo-Re, Mo-Ru, Nb-Re and Nb-Ru, where the μ and Laves phases became more and more stable as compared with A15 and σ as the size-factor increased in going from Mo-Re to Nb-Ru.

In other studies[81], a combination of density functional theory, tight-binding models and analytical bond-order potentials was used to investigate structural trends in TM_5Si_3 compounds, where TM was titanium, zirconium, hafnium, vanadium, niobium, tantalum, chromium, molybdenum or tungsten. The formation energies of competing structure types were first calculated by using density functional theory. In agreement with experimental results, density functional theory predicted a D88 to D8m structural trend across the 3d series and a D88 to D8l to D8m trend across the 4d and 5d series. A p-d tight-binding model reproduced these trends and increased confidence in the application of bond-order potential theory. The application of moment-analysis to the latter results indicated that the density-of-states up to the fifth moment was required in order to explain the structural trends across the 3d series. Up to the ninth moment was required in the case of the 4d and 5d series.

It was recalled that, while group-VII 4d technetium and 5d rhenium have hexagonally close-packed ground states, 3d manganese exhibits a complex χ-phase ground state with a complex non co-linear magnetic ordering. Density functional theory calculations showed[82] that, without magnetism, the χ-phase remained the ground state of manganese; thus implying that magnetism and the atomic-size difference between large- and small-moment atoms were not critical factors influencing the anomalous stability of the χ-phase with respect to hexagonal close-packing. A tight-binding model showed that, with a more than half-filled d band, harder potentials stabilized close-packing while a softer potential stabilized the more open χ-phase. By analogy with the structural trend from open to close-packed phases in going down the group-IV column, the anomalous stability of the χ-phase of manganese was attributed to the 3d-valent manganese lacking d states in the core. This then led to a softer repulsion between atoms than that in 4d technetium and 5d rhenium. An analytical bond-order potential was further used to investigate the structure of manganese at 0K. The bond-order potential qualitatively reproduced the binding-energy curves of α-, β-, γ-, δ- and ε-Mn. It also predicted the complex co-linear antiferromagnetic ordering in α-Mn, the ferrimagnetic ordering in β-Mn and the antiferromagnetic ordering in γ-, δ- and ε-Mn that were predicted by density functional theory. A bond-order potential expansion which included 14 moments was sufficient to reproduce most of the properties indicated by the tight-bonding model, apart from the elastic shear constants (table 2). Those required higher moments. It has been shown[83] that a low-dimensional moment-descriptor can be sufficient. This is because the lowest moments, calculated on the basis of the most propinquitous atomic neighborhood, make the largest contributions to the local bond energy. Moment-descriptors were constructed which projected the space of local atomic environments onto a 2-dimensional map. Distances in the map could be related to the energy differences between local atomic environments.

Other investigators essayed a multitude of other approaches. Watson and Bennett[84] considered that the changes in volume which were associated with alloying were important in governing the mechanical properties of alloys. Correlations were noted which revealed a general influence of volume effects, with some exceptions. A parameter was also introduced which resembled an electronegativity (figure 11). The most notable deviations from the general trend for volume contractions in transition-metal compounds involved the occurrence of magnetism. There was, for example, the tendency of systems to exhibit a body-centered cubic structure where - in the absence of magnetism - they would be expected to adopt a close-packed structure.

*Table 2. Elastic constants of manganese, calculated using
tight-binding and 20- or 14-moment bond-order potentials*

Allotrope	Magnetic State	Method	B_0 (GPa)	C' (GPa)	C_{44} (GPa)
α	non-magnetic	TB	237	57	41
α	non-magnetic	BOP(20)	257	55	37
α	non-magnetic	BOP(14)	235	47	30
α	antiferromagnetic	TB	163	54	68
α	antiferromagnetic	BOP(20)	193	73	85
α	antiferromagnetic	BOP(14)	152	72	77
β	non-magnetic	TB	256	40	78
β	non-magnetic	BOP(20)	256	37	86
β	non-magnetic	BOP(14)	259	21	70
β	ferrimagnetic	TB	218	49	78
β	ferrimagnetic	BOP(20)	220	51	86
β	ferrimagnetic	BOP(14)	210	39	74
γ	non-magnetic	TB	237	43	82
γ	non-magnetic	BOP(20)	240	10	70
γ	non-magnetic	BOP(14)	241	17	107
γ	antiferromagnetic	TB	166	64	80
γ	antiferromagnetic	BOP(20)	179	77	101
γ	antiferromagnetic	BOP(14)	189	75	110
δ	non-magnetic	TB	248	-65	18
δ	non-magnetic	BOP(20)	250	-45	-32
δ	non-magnetic	BOP(14)	248	-42	4
δ	type-2 antiferromagnetic	TB	72	-1	42
δ	type-2 antiferromagnetic	BOP(20)	111	1	18
δ	type-2 antiferromagnetic	BOP(14)	97	6	41

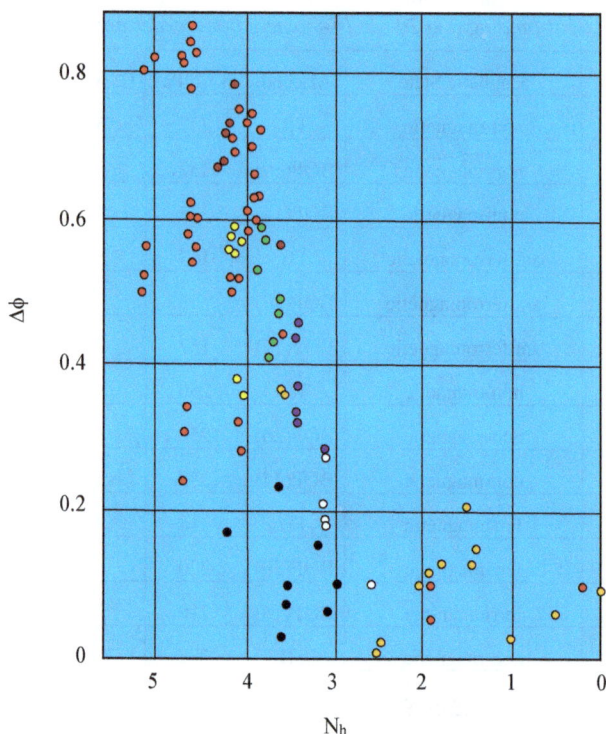

Figure 11. Structures of intermetallic compounds, predicted on the basis of electronegativity-difference and average d-band hole count (number of unoccupied levels per atom below the top of the d band). Red: B2, yellow: orthorhombic, green: B33, brown: B27, orange: $L1_0$, white: B19, purple: B11, black: D8b

A striking deviation from the general volume trend occurred in the case of alloys of noble metals with lanthanum and neighboring lanthanides such as cerium and praseodymium. The volume contractions of these systems were much lower than those associated with the same light lanthanides when alloyed with transition elements. In the case of noble and transition metals, d-band hybridization played an important role in alloy formation. Starting with the Cr-Mo-W column of the periodic table and moving leftwards, it was noted that electronegativities decrease, atomic volumes and compressibilities increase; as do volume contractions. A break in volume effects occurred when zinc, cadmium and mercury were involved and, as with the noble metals, the volume contractions which were associated with their alloys with lanthanum were anomalously small. The main trend in volume effects was followed so well by transition and noble metals that it could be used to predict volume effects. The volume trend held well for 3:1 systems.

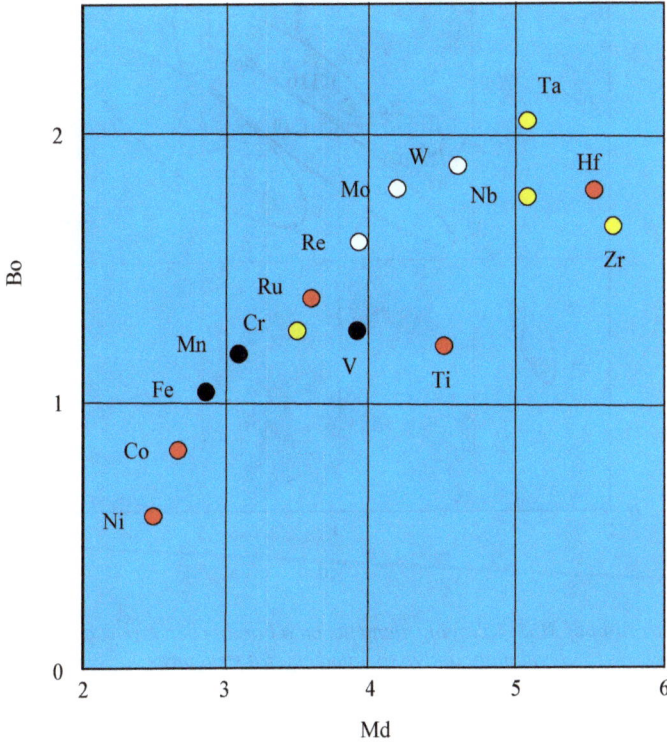

Figure 12. Location of MoSi₂ alloying elements on a bond-order versus d-orbital energy-level map, and associated solid solubilities, Red: insoluble, black: unknown, yellow: partially soluble, white: completely soluble

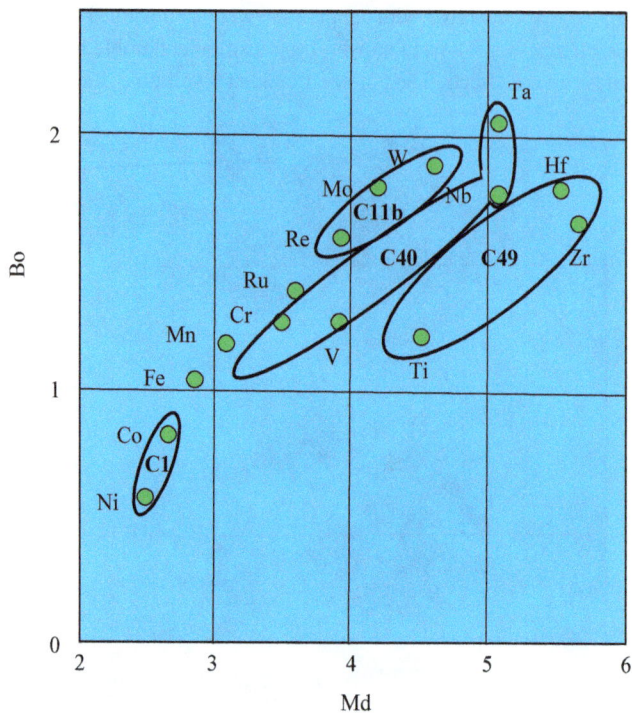

Figure 13. Location of MoSi₂ alloying elements on a bond-order versus d-orbital energy-level map, and associated crystal structures

Others found[85] no correlation between estimated stability differences and the positions of points on the above Watson-Bennett structure map, although it was noted that unstable compounds could be divided into two groups: ones with a small (850J/mol) stability difference with respect to stable states and ones with a large (2100J/mol) stability difference.

Materials Research Forum LLC
https://doi.org/10.21741/9781644902011

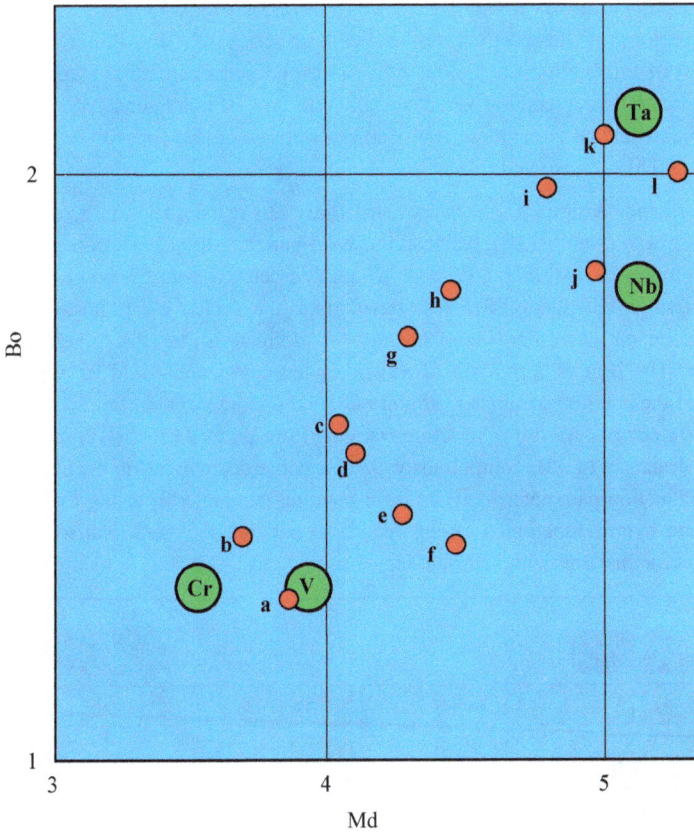

Figure 14. Enlargement of the C40-structure area of the previous figure, showing silicide locations deduced from the calculations: $a - Cr_{1-x}Si_2Ti_x$ (x = 0 to 0.6), $b - Cr_{1-x}Mo_xSi_2$ (x = 0 to 0.29), c - $Re_9Si_{20}Ti$, d - $Mo_xV_{1-x}Si_2$ (x = 0 to 0.5), $e - ReSi_{10}Ti_4$, $f - Nb_xSi_2Ta_{1-x}$ (x = 0 to 1), $g - Mo_xTi_{1-x}Si_2$ (x = 0.29 to 0.85), $h - Nb_{1-x}Mo_xSi_2$ (x = 0 to 0.81), $i - Ta_{1-x}W_xSi_2$ (x = 0.8 to 1), $j - W_{1-x}Nb_xSi_2$ (x = 0.713 to 1), $k - Ta_{1-x}W_xSi_2$ (x = 0 to 0.2), $l - Ta_{1-x}Zr_xSi_2$ (x = 0 to 0.2)

It was recognized early on[86] that the maps could reflect general trends in segregation and order at low temperatures in binary substitutional transition-metal alloys. The occurrence of ordered structures at 0K, for a given lattice, could be deduced from the electronic structure using perturbation methods for concentrations of 19, 15, 14, 13 and 12%. The stability regions of these structures, defined in terms of band-filling and diagonal disorder, were related to those regions found in experimental structural maps. Most of the observed trends in segregation versus order, or in type of homogeneous order, for face-centered cubic and body-centered cubic lattices were accounted for with regard to Ni_8Nb, Ni_4Mo, $L1_2$, $D0_{22}$, Pt_2Mo and

B2 long-period antiphase boundary structures. In a similar spirit, an empirical relationship was established[87], between the band-gap and orbital electronegativity in *sp*-bonded compounds, by using a formula which was derived from the bond orbital model. The abscissa in the relationship was based upon the orbital electronegativity and the average principal quantum number and was defined to be the so-called bonding parameter. The heat of formation of an *sp*-bonded compound could be expressed as a function of the bonding parameter, orbital radii and the ratio of atomic fraction to number of valence electrons. Maps for structures ranging from B1 to B4 for AB compounds and 9 different structures of AB_2 compounds were also constructed by using the hybrid function and gap reduction parameter; derived from bonding parameter by using the bond orbital model. The hybrid function and gap reduction parameter were found to be excellent descriptors for characterizing the bonding of compounds.

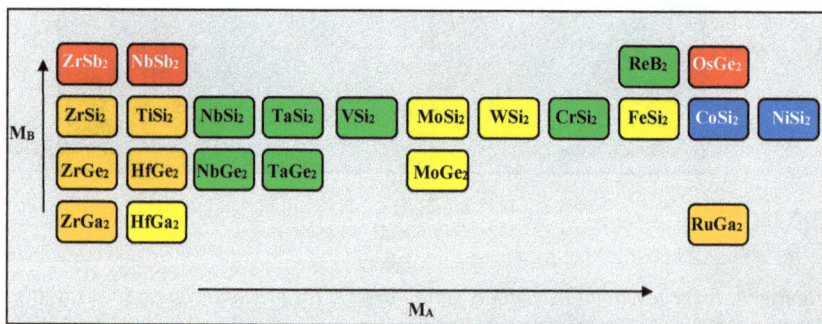

Figure 15. The AB_2 structure-map, based upon Mendeleev numbers, around $MoSi_2$. Red: monoclinic, orange: orthorhombic, yellow: tetrahedral, green: hexagonal, blue: cubic

A modification of the Hückel tight-binding calculation was applied[88] to intermetallics of the form, ZA_3 or ZA_6, where Z was a more electropositive atom (groups I to V, or a lanthanide) and A was a more electronegative atom (late transition-metal, or main-group element). The main covalent bonds in these compounds were those between the A-atoms.

The electron concentration was calculated by using the Zintl concept: that is, the more electropositive element was assumed to donate its valence electrons to the covalent framework of A-atoms. The total energy was then calculated only for the covalent network of A-atoms. Second-moment scaled Hückel energies were subsequently used to construct structure maps for these intermetallic compounds by plotting differences in the total energy as a function of the number of electrons per atom. Such maps could correctly demarcate the zones of stability occupied by structures of the types: $AuCu_3$, $TiNi_3$, $TiCu_3$, BiF_3, $SnNi_3$, $NdTe_3$, TiS_3, $SmAu_6$, $CeCu_6$ and $PuGa_6$.

A simple model was suggested[89] for predicting the structural stability of AB intermetallic compounds, and the relative stability of the 4 most common types of atomic environment were calculated by using the tight-binding model. Calculated 3-dimensional structure maps, involving the difference in the valence electron orbital energy of an atom, the distance between atoms and the average number of electron per atoms, exhibited good agreement with a corresponding semi-empirical quantum theory-based diagram. The calculations showed that the electronic factor, size factor and valence electron factor were very important in determining the type of environment. For an AB compound with lower valence number, a system with a large difference in atomic energy level and large difference in atomic size preferred the CN6 type of atomic environment while a system with a small difference in energy level and small difference in atomic size preferred the CN12 or CN14 types of environment. It was predicted that CN4 was much less likely to occur in systems with a lower valence number.

Alloying effects upon the electronic structure of $MoSi_2$ were investigated[90] by using molecular orbital methods. Strong covalent Si-Si and Mo-Si bonds existed in the $MoSi_2$, and the nature of the chemical bonds was modified by alloying. The bond-order parameter and d-orbital energy level were determined for various alloying elements, and the solid solubilities of the alloying elements could be described in terms of those parameters (figure 12). A structure-map (figure 13) of transition-metal disilicides was constructed by plotting bond-order against the orbital energy level. Compounds were clearly separated on the map, according to crystal structure, and those having the same crystal structure gathered together. The $MoSi_2$ compound, with its $C11_b$-type structure was located in a high-Bo region of the map. This implied that it had stronger chemical bonds than other compounds, and this explained its high melting-point. There was a C40-type region near to the $C11_b$-type region and this was reasonable because the C40 structure, albeit as a metastable phase, was to be found following rapid quenching from the melt. The introduction of a stacking fault into the (110) plane of a $C11_b$ structure also created identical atomic arrangements to those existing on the (00•1) plane in the C40 structure. Structural modification into a C40 compound was therefore possible by adding elements such as chromium, niobium and tantalum (figure 14). Iron, cobalt, nickel, chromium, vanadium, titanium and niobium were chosen[91] as ternary alloying elements according to an AB_2 structure map or atomic size factor. The iron, cobalt and nickel exhibited no solid solubility in as-cast $MoSi_2$, while chromium, vanadium, titanium and niobium exhibited a limited solid solubility: 1.4, 1.4, 0.4 and 0.8, respectively. By using the Mendeleev numbers, an approximate structure-map (figure 15) could be arranged around $MoSi_2$.

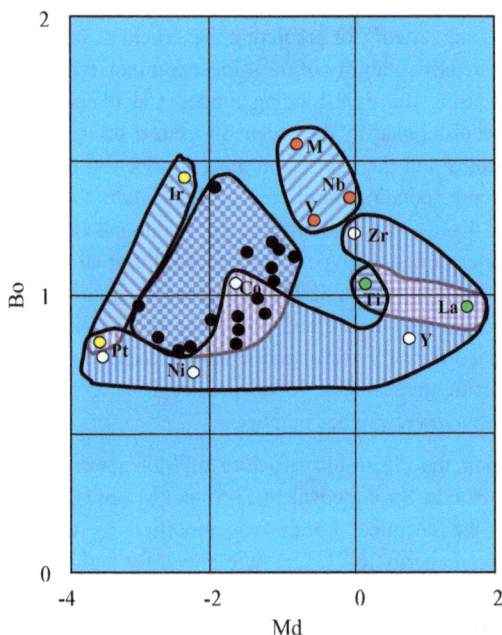

Figure 16. Structure-map for M_3Al compounds, based upon bond-order and difference in energy of d-orbitals. White (vertical hatching): $L1_2$, red (upward-diagonal): $A15$, green (horizontal hatching): $D0_{19}$, yellow (downward-diagonal): $GaPt_3$, black (chequered): $L2_1$. Unlabelled black points are compounds involving 3 elements; with 2 in the form, A_2B.

Other structure-maps were plotted[92], on the basis of bond-order and d-orbital energy level, for aluminides (figures 16 and 17), silicides and transition-metal based compounds (figures 18 and 19). There was again a clear demarcation of the crystal structures. The Bo and Md values were calculated for an octahedral M_2Al_4 cluster. The crystal structures are well separated in each map. On the M_3Al map, the A15-type region is located in the high-

Bo region and the DO_{19}-type region is located between the A15- and $L1_2$-type regions. Ternary compounds can have either a DO_3- or $L2_1$-type structure. These resemble one another, and their regions almost overlap on the map. On the MAl_3 map, there is a DO_{22} region near to the DO_{23} region because of their mutual structural resemblance ... to such an extent that $HfAl_3$ can adopt either structure, depending upon the temperature; as shown by the orange and red circles on the map. On the M_3Ti structure-map, Al_3Ti is located in the highest-Md region and the $L1_2$ covers a lower-Md region than does DO_{22}; suggesting that substituting lower-Md elements for aluminium atoms benefits structural modification, and it is known that the DO_{22}-type structure is converted to $L1_2$ by adding vanadium, chromium, manganese, iron, nickel or copper to Al_3Ti. Factors which influence the deformation mechanisms of Al_3Ti and Al_3V had been studied[93] elsewhere, showing that plastic deformation of the former involved micro-twinning and the glide of partial dislocations with a Burgers vector of $\frac{1}{2}<110>$ moving on (001) planes and producing ribbons of antiphase domain boundary. In the case of Al_3V, these mechanisms were not activated and deformation occurred mainly via the glide of dislocations with a Burgers vector of $\frac{1}{2}<110]$ moving on $(1\bar{1}2)$ planes. This difference in behavior was attributed to the relative stability of the DO_{22} crystal structure as compared with the Cu_3Ti-type structure, where micro-twinning operates, and DO_{23}, where antiphase domain boundaries on basal planes operate. It was further suggested that an increased ductility of these compounds could be achieved by alloying with elements which reduce the energy-difference between their DO_{22} and Cu_3Ti-type crystal structures. An electronic model for the DO_{22}-to-$L1_2$ phase transformation in group-IVA tri-aluminides when lightly alloyed with ternary transition-metal elements had also been proposed[94]. Those transition-metal substitutions which provoked phase-change were those which imparted d-characteristics to the aluminium sp-bonding region. This in turn was thought preferentially to order aluminium second-neighbor atoms; a preferential order which stabilized the $L1_2$ structure. This was part of a philosophy which held[95] that the structure of intermetallic alloys was then interpreted too narrowly and that, by extending the definition of structure so as to include aspects of the electronic charge distribution, relationships might be found to exist between intrinsic mechanical properties and the extended definition of structure. The definition of this extended structure required only a knowledge of the critical points of the total charge density. The sought-for relationships between intrinsic mechanical properties and the extended definition of structure then originated from the nature of the charge redistribution which was associated with strain. The direction of the charge redistribution was governed only by that extended structure, while its degree was found by quantifying the latter. This had been demonstrated[96] by determining the extended structure and nature of the charge redistribution that resulted from the uniaxial straining of alloys, CuAu and TiAl, which had a $L1_0$ structure. Those alloys had the same crystallographic structure, but their extended structures were different: CuAu had the same extended structure as that of the allotropic face-centered cubic metals, but TiAl did not. The differing extended structures gave rise to differing charge redistributions, and these were thought to be related to the intrinsically ductile behavior of CuAu and the tendency to transgranular failure of TiAl. The concept of differing electronic structures had apparently been inspired by even earlier work[97] in which representative wave-functions had been determined for copper, Ir_3Cr and Zr_3Al using multiple-scattering cluster calculations. It had been noted that, although parameters such as s-orbital electronegativity predicted an isotropic charge density for Ir_3Cr and an anisotropic charge density for Zr_3Al, entirely different characteristics were in fact observed. This was explained in terms of an effective s-orbital electronegativity which was sensitive to the local atomic surroundings. The charge density which actually resulted from the effective electronegativities could be closely linked to observed mechanical properties, with the most isotropic charge density leading to the most ductile material.

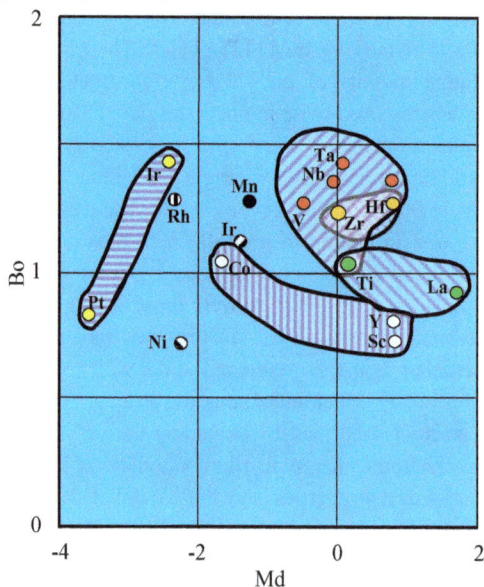

Figure 17. Structure-map for MAl₃ compounds, based upon bond-order and difference in energy of d-orbitals. White (vertical hatching): L1₂, red (upward-diagonal hatching): D0₂₂, green (downward-diagonal hatching): D0₁₉, yellow (horizontal hatching): AsNa₃, orange (chequered): D0₂₃. Black circle: Al₃Mn, vertically-hatched circle: D0₂₄, upward-diagonal hatched circle: D2h²³, downward-diagonal hatched circle: D0₁₁.

In the case of the MNb_3 structure map there are many A15-type compounds, but the $L1_2$-type structure is limited to Nb_3Si alone. The ternary compound, $Nb_3(Al,Si)$, conserves an A15-type structure; structural modification of Nb_3Al, when third elements are substituted for aluminium, is expected to be difficult.

By combining structure-controlling factors for compounds of the form, M_2Q, where M was a transition metal that was poor in valence electrons and Q was a pnicogen or chalcogen, the metal-rich pnictides and chalcogenides of groups 3, 4 and 5, which can exhibit 11 different

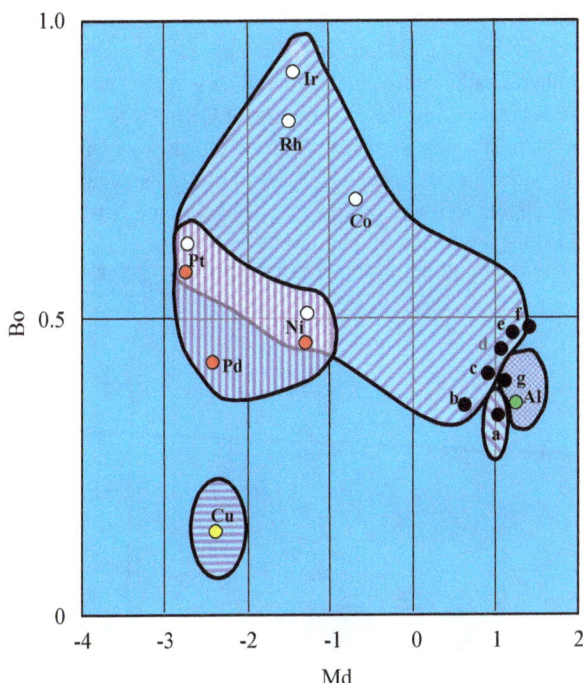

Figure 18. Structure-map for M_3Ti compounds, based upon bond-order and difference in energy of d-orbitals. White (upward-diagonal hatching): $L1_2$, red (vertical hatching): $D0_{24}$, green (chequered): $D0_{22}$, yellow (horizontal hatching): $D0_a$, downward-diagonal hatching: $D0_{23}$. a: $Al_{71}Cu_4Ti_{25}$, b: Al_5CuTi_2, c: $Al_{67}Ni_8Ti_{25}$, d: $Al_{22}Fe_3Ti_8$, e: $Al_{67}Mn_8Ti_{25}$, f: $Al_{67}Cr_8Ti_{26}$, g: $(Al,Si)_3Ti$.

structures, could be plotted[98] onto a 2-dimensional structure-map which clearly demarcated their domains. The map had a combination of atomic factors, such as principal quantum numbers and radii as the abscissa, and the averaged coordination number of the Q-atoms as the ordinate[99]. It was noted[100] that ZrTiAs had an La$_2$Sb-type structure, in perfect agreement with the structure-map. Although Zr$_2$As was predicted to have a Zr$_2$P-type structure, it had not yet been prepared. The ZrTiAs crystallized with tetragonal space-group I4/mmm symmetry, and with a = 379.29pm, c = 1480.2pm and V = 212.95 x 10^{6}pm^3 (Z = 4). It exhibited complete ordering of the zirconium and titanium atoms, but the zirconium atom could be partly replaced by titanium, giving Zr$_{1-x}$Ti$_{1+x}$As (x = 0 to 0.42). The structure contained strong Zr-As, Ti-As and Ti-Ti bonds, plus some weaker Zr-Zr and Zr-Ti bonding interactions. Among 28 of investigated[101] RCuZn, RAgZn and RAgAl intermetallic compounds, where R was a rare earth, YbAgAl was of MgZn$_2$ structure-type whereas all of the others were of CeCu$_2$ type. A familiar decrease in the unit-cell parameters in going from lanthanum to erbium was observed, and a clear positive deviation of 3 ytterbium compounds was attributed to the divalency of ytterbium. The use of structure-mapping to RMX compounds, where M and X were elements in the 10 to 15 groups, guided the prediction of then-unknown phases.

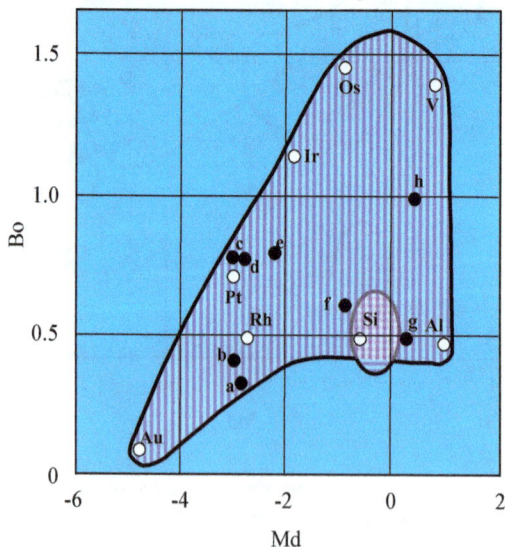

Figure 19. Structure-map for MNb$_3$ compounds, based upon bond-order and difference in energy of d-orbitals. White (vertical hatching): A15, horizontal hatching: L1$_2$. a: Nb$_6$PdRh, b: Nb$_6$PdPt, c: AuOsNb$_6$, d: (Fe,Pt)$_{24}$Nb$_6$, e: Nb$_6$RhRu, f: IrSi$_4$Nb$_{15}$, g: Nb$_3$(Al,Si), h: (Al,Cr)Nb$_3$.

The relationship between Mendeleev numbers and the type of phase diagram was considered[102] for group-VIII and group-IB group elements as solvents and other elements as solutes; the Mendeleev numbers of the solvents and solutes being expressed as M_A and M_B, respectively. A 2-dimensional M_A versus M_B map exhibited an oblique line which divided the types of phase diagram of the solvent metals into two symmetrical parts: those of the other elements distributed above or below the line. The phase diagrams between the solvent metals were generally simple systems: mainly complete solid solution or peritectic, with some 40% of each type. The solvent metals could be divided into 3 groups: cobalt, iridium, rhodium, nickel, platinum, palladium | silver, gold, copper | iron, osmium, ruthenium. Some 80% of the phase diagrams were complex and fewer than 20% were simple.

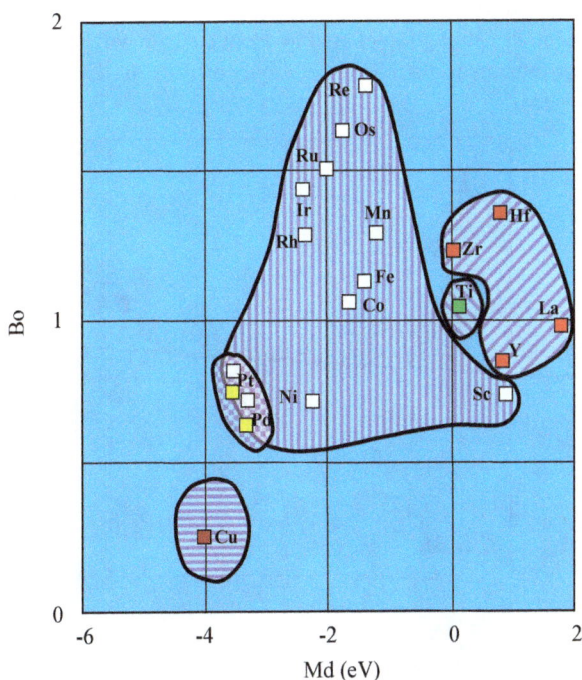

Figure 20. Structure-map for MAl compounds, based upon bond-order and d-orbital energy level. White (vertical hatching): B2, red (upward-diagonal hatching): B33, green (downward-diagonal hatching): L1₀, yellow (chequered): B20, orange (horizontal hatching): B8₁. Note that the platinum and palladium compounds both fall into different groups, depending upon the temperature.

The type of atomic environment, coordination polyhedra, which was experienced by each element in a binary compound having an equi-atomic composition was analyzed[103] on the basis of a comprehensive set of data. The Mendeleev number was found to classify the environments in a useful way. A map which plotted the maximum Mendeleev number versus the ratio of the minimum to maximum Mendeleev number sub-divided the systems into distinct stability domains. Note however that the Mendeleev numbers used here differed from those pioneered by Pettifor. The maps also clearly demarcated those systems, where intermediate compounds formed, from those where no compounds formed. This made it possible to predict the existence of unknown compounds having a particular atomic environment. It was found that, overall, the elements preferred to adopt a limited number of atomic environments: tetrahedron, trigonal prism, octahedron, cube, tricapped trigonal prism, cuboctahedron, rhombic dodecahedron, heptacapped pentagonal prism. Each element also strongly preferred to adopt a single polyhedron within a given phase. A highly symmetrical distribution of atoms of differing kinds existed in mixed coordination polyhedra.

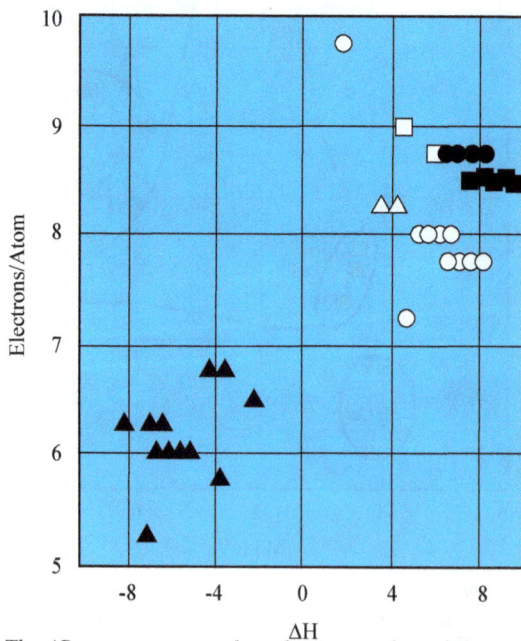

Figure 21. The AB₃ structure-map, based upon number of electrons per atom and difference in energy of d-orbitals. Open circles: AuCu₃, open squares: TiCu₃, open triangles: SnNi₃, filled circles: TiAl₃, filled squares: TiNi₃, filled triangles: Cr₃Si.

New pnictides, HfTiP, HfTiAs and HfVAs, were prepared[104] which had the predicted distorted-$TiAs_4$ and VAs_4 tetrahedral structures, while the structures of the zirconium analogues, ZrTiAs and ZrVAs, consisted of $TiAs_4$ and VAs_4 square planes, respectively. The differences were attributed to marked differences in the metal-atom sub-structures. The HfTiP and ZrTiP materials were isostructural. These variations were predicted by the above structure-map for metal-rich pnictides and chalcogenides. The HfMQ structure was again stabilized by strong Hf-Q, M-Q and M-M bonds and by weaker Hf-Hf and Hf-M interactions. The 3 new $Hf_{1-x}M_{1+x}Q$ phases had wide ranges.

	Y	Sc	Zr	Hf	Ti	Tc	Re	Os	Ru	Co	Mg	Cd	Zn	Be	Tl
Y															
Sc	•														
Zr	•	•													
Hf	•	•	•												
Ti	•	•	•	•											
Tc	•	•	•	•	•										
Re	•	•	•	•	•	•									
Os	•	•	•	•	•	•	•								
Ru	•	•	•	•	•	•	•	•							
Co	•	•	•	•	•	•	•	•	•						
Mg	•	•	•	•	•	•	•	•	•	•					
Cd	•	•	•	•	•	•	•	•	•	•	•				
Zn	•	•	•	•	•	•	•	•	•	•	•	•			
Be	•	•	•	•	•	•	•	•	•	•	•	•	•		
Tl	•	•	•	•	•	•	•	•	•	•	•	•	•	•	

Figure 22. High-throughput prediction of compound-formation in binary alloys of hexagonal close-packed metals. Light blue: phase-separating, purple: compound-forming, red: phase-separation observed but stable ordered HCP or non-HCP structure predicted, green: phase-separation observed but stable ordered HCP structure predicted, black: phase-separation observed but stable ordered non-HCP structure predicted

Crystal structure-maps were constructed[105] for the prediction of the structures of intermetallic compounds (figure 20), and the phase stability of the metal hydrides which formed in hydrogen-storage compounds was deduced from the nature of the chemical bonds between the atoms. When an alloying element was added, the chemical interactions between the atoms naturally varied in a manner which depended upon the addition. This was modeled in terms of the chemical bond between atoms in a small metal polyhedron, where the hydrogen was located at the center, and structural changes during hydrogenation. When the crystal structure of the hydride was a derivative of the original one, as in the case of $LaNi_5$, TiFe and $ZrMn_2$, the hydrogen atom occupied interstitial octahedral or tetrahedral sites; causing lattice expansion and distortion during hydrogenation. The ease of lattice expansion and distortion during hydrogenation then affected hydride stability. This was controlled mainly by the type of chemical bonding between the metal atoms in the initial compound, rather than by the degree of direct metal-hydrogen interaction. When the crystal structure of the hydride was an entirely novel one, such as Mg_2Ni, metal-hydrogen interaction was important with regard to hydride stability. In the case of body-centered cubic vanadium, only the metal-hydrogen interaction was responsible for the observed stability of VH_2.

	Y	Sc	Zr	Hf	Ti	Tc	Re	Os	Ru	Co	Mg	Cd	Zn	Be	Tl
Y															
Sc															
Zr															
Hf															
Ti															
Tc			●	○	○										
Re					●										
Os			○	○	○	●	●								
Ru		○	○	○	○	●	●	●							
Co	●	○	●	●	○	●	●								
Mg	○	●	●	●		●			●	●					
Cd	○	○	●	●	●						●				
Zn	○	○	○		●										
Be				●			●		○						
Tl	○	●									●				

Figure 23. High-throughput prediction of AB structures in alloys of hexagonal close-packed metals. Light blue: B2, purple: B19, red: phase-separation observed but ordered structure predicted, green: CdTi, black: discrepancy between observed and predicted structure. A horizontal, B vertical.

A 2-dimensional structure-map for AB_3 transition-metal compounds was based[106] upon plotting the electron count versus the difference in d-orbital Coulombic integrals. The map (figure 21) could demarcate the 6 then-known AB_3 transition-metal structures: Cr_3Si, $AuCu_3$, $SnNi_3$, $TiAl_3$, $TiCu_3$ and $TiNi_3$. The main structural features which led to energy differences were the various numbers of 3- and 4-membered rings of bonded atoms. The plotting of 35 known phases onto the map showed that the icosahedral Cr_3Si structure was taken up by systems which were very chemically different to close-packed structures. The Cr_3Si structure was exhibited by systems 5 to 7 electrons/atom, where the minority component was rather more electronegative than was the majority component. The closest-packing structures were found for systems with 7 to 10 electrons/atom, where the A-atom was more electropositive than was the B-atom. The $AuCu_3$ structure was preferred in 2 regions: one where ΔH_{ii} was close to 0 and one where it was between 4 and 9eV and there were 8.5 or fewer electrons/atom. In this region, as the number of valence electrons increased beyond 8 per atom, the $SnNi_3$ structure appeared and then the $TiNi_3$ structure. At 8.75 electrons/atom, the $TiAl_3$ and $TiCu_3$ structures were observed.

Table 3. Relative contributions of various factors to the classification of AB_2 structures. Total = $aVE + bSZ + cEC$, where VE is the valence-electron effect, SZ is the size-effect factor, EC is the electrochemical factor and $a + b + c$ is necessarily equal to unity

a	b	c	Structure-Type
0.56642	0.28488	0.14870	AlB_2
0.41081	0.31347	0.27572	$CaIn_2$
0.42397	0.37525	0.20078	$CuAl_2$
0.47176	0.30104	0.22720	$HfGa_2$
0.59414	0.25564	0.15022	$MgCu_2$
0.51994	0.34564	0.13442	$MgZn_2$
0.501 56	0.29063	0.20781	$MoPt_2$
0.64461	0.35539	0.00000	$OsGe_2$
0.60220	0.22398	0.17382	$ZrGa_2$
0.50044	0.12606	0.37350	CaC_2
0.57268	0.16554	0.26178	CaF_2
0.50315	0.18305	0.31380	Cd_2Ce
0.60343	0.09648	0.30009	$CoSb_2$

0.66730	0.08762	0.24508	$CrSi_2$
0.61640	0.15175	0.23185	FeS_2
0.60809	0.04424	0.34767	Hg_2U
0.74794	0.08445	0.16761	$HoSb_2$
0.64570	0.17107	0.18323	KHg_2
0.74794	0.08445	0.16761	$LaSb_2$
0.68852	0.10321	0.20827	Mg_2Cu
0.73371	0.06269	0.20361	$MoSi_2$
0.74794	0.08445	0.16761	$NdAs_2$
0.63446	0.14181	0.22373	$ThSi_2$
0.56374	0.09609	0.34017	$TiSi_2$
0.70688	0.09323	0.19989	$ZrSi_2$
0.07753	0.72511	0.19736	Co_2P
0.08043	0.61518	0.30439	Cu_2Sb
0.32216	0.36067	0.31717	$CuZr_2$
0.29069	0.43908	0.27023	Fe_2P
0.17337	0.41551	0.41112	Ni_2In
0.13926	0.51980	0.34094	Ti_2Ni
0.00000	0.40165	0.59835	La_2Sb
0.18575	0.27444	0.53981	$PbCl_2$

Structure-maps (figures 22 to 25) for binary alloys of hexagonal close-packed metals were prepared[107], using data-mining, for poorly-understood systems which were believed to be of phase-separating type. These maps showed that the clusters of non compound-forming systems were considerably smaller than was suspected. Among the 105 binary metallic systems which were considered, 46 had previously been reported as phase-separating. Calculations showed that 18 of those involved compounds which formed at low temperatures. Most of the predicted compounds were not derived from hexagonal close-packed structures. Some alloys were reported to be phase-separating, but were predicted to have stable ordered structures. No cases were found where a system was reported to be compound-forming but was predicted to be phase-separating at low temperatures. Most of the 46 reported phase-separating systems were grouped, in a map based upon the Pettifor scale, into 3 clusters. Thallium was implicated in 10 phase-separating systems, 7 of which were not located in those 3 clusters.

In what might be termed a meta-level approach to structure-maps, attention was paid[108] to the collective effect of the various chemical and bonding parameters, which are assumed to govern the stability of compounds, by exploiting information-entropy principles. It was then possible to establish a quantitative scaling parameter, in Shannon-entropy parlance, that mapped the relative contributions of the parameters onto a single map. This map resembled those which are used in colorimetry and voting analysis, in that a point plotted in an equilateral triangle showed the relative proportional effect (table 3) of each parameter when applied to an AB_2 intermetallics database.

	Y	Sc	Zr	Hf	Ti	Tc	Re	Os	Ru	Co	Mg	Cd	Zn	Be	Tl
Y										●	●	●	●		
Sc			●		●		●	●	●	●	●	●			●
Zr				●	●				●			●	●		
Hf					●		●		●			●	●		●
Ti				●	●	●	●	●	●	●		●	●		
Tc	●	●	●	●							●		●	●	
Re	●	●	●	●				●						●	
Os	●	●	●	●											
Ru	●	●	●	●									●		
Co	●	●	●	●	●						●				
Mg	●												●		●
Cd	●		●												
Zn	●	●	●	●	●						●				
Be			●	●	●	●	●		●	●			●		
Tl															

Figure 24. High-throughput prediction of AB_2 structures in alloys of hexagonal close-packed metals. Blue: CuZr$_2$, red: phase-separation observed but ordered structure predicted, green: C14, black: discrepancy between observed and predicted structure, orange: C15. A horizontal, B vertical.

In recent years, the Mendeleev number has proved useful in the more direct search for desirable mechanical properties, such as superhardness. Empirical guides to Vickers hardness and fracture toughness have permitted the rapid screening of materials in the search for optimum properties[109]. Such methods have similarly proved to be useful in studying so-called MAX phases. These are layered hexagonally-structured ternary carbides or nitrides involving a transition-metal and an A-group element. They exhibit a surprising combination of metallic and ceramic characteristics. As usual nowadays, the key strategy is to create a prediction-system which incorporates all of the relevant databases as well as generally useful heuristic principles. Structure-mapping here[110] again proves its value, with the classic Hume-Rothery parameters guiding the search for new MAX phases. Formable and non-formable compositions could be demarcated on a 2-dimensional plot having geometrical and electron-concentration factors as coordinates.

	Y	Sc	Zr	Hf	Ti	Tc	Re	Os	Ru	Co	Mg	Cd	Zn	Be	Tl
Y						•		•	•	•					
Sc						•		•		•					•
Zr		•													•
Hf		•				•	•								
Ti						•	•	•	•				•		•
Tc						•	•	•	•						
Re								•	•	•					
Os						•	•		•						
Ru						•	•	•					•		
Co	•				•										
Mg	•	•							•			•			•
Cd	•	•	•								•				
Zn	•	•	•	•	•	•				•					
Be					•	•		•	•	•					
Tl	•														

Figure 25. High-throughput prediction of AB₃ structures in alloys of hexagonal close-packed metals. Blue: $D0_{19}$, red: phase-separation observed but ordered structure predicted, dark green: $D0_{11}$, black: discrepancy between observed and predicted structure, orange: A15, purple: $L1_2$, yellow: YZn_3, brown: Cd_3Er, light green: $PuNi_3$. A horizontal, B vertical.

The Advent of Data-Mining

Online sources such as Pearson's Crystal Data now offer data on the structures of tens of thousands of intermetallic phases. The characteristics of these are of particular interest because they can often be 'usefully anomalous' as when, for example, the tensile strength increases with temperature. Stable quasicrystals and high-T_c superconductors may similarly exhibit unusual physical properties which are related to characteristic crystal structures. High-temperature ferroelectrics and superconductors exhibited strong localization on quantum-structure diagrams which were constructed[111] so as to organize the complete database of intermetallic compounds according to their local structure. There was also a strong diagrammatic relationship between the ferroelectrics and superconductors, thus reinforcing the concept that both phenomena involved factors which might be linked to lattice instability and anharmonicity.

	Os	Ru	Ir	Rh	Pt	Pd
Zn	●	●	●	●	●	●
Cd	●	●	●	●	●	●
Hg	●	●	●	●	●	●
Cu	●	●	●	●	●	●
Ag	●	●	●	●	●	●
Au	●	●	●	●	●	●
Pd	●	●	●	●	●	
Pt	●	●	●	●		●
Ni	●	●	●	●	●	●
Rh	●	●	●		●	●
Ir	●	●		●	●	●
Co	●	●	●	●	●	●
Ru	●		●	●	●	●
Os		●	●	●	●	●
Fe	●	●	●	●	●	●
Mn	●	●	●	●	●	●
Re	●	●	●	●	●	●
Tc	●	●	●	●	●	●
Cr	●	●	●	●	●	●
W	●	●	●	●	●	●
Mo	●	●	●	●	●	●
V	●	●	●	●	●	●

Ta	●	●	●	●	●	●
Nb	●	●	●	●	●	●
Ti	●	●	●	●	●	●
Hf	●	●	●	●	●	●
Zr	●	●	●	●	●	●
Sc	●	●	●	●	●	●
Y	●	●	●	●	●	●

Figure 26. Prediction of compound-formation in alloys of platinum-group metals. Green: compound formed as predicted, black: no compound formed as predicted, orange: disordered phases observed but low-energy compounds predicted, blue: no low-temperature stable compounds observed but disordered phase predicted, red: no compound observed but low-temperature compounds predicted

Systematically testing of all of the myriad intermetallics in search of such anomalous properties is not economically feasible; hence the interest in correlating properties with structure in a theoretical manner, as already described here. This has always been done of course, but sophisticated computer techniques now furnish tools such as pattern-recognition, genetic algorithms and artificial intelligence which speed up the process and create shortlists of materials which may be of particular interest for detailed testing.

Nearly four decades ago an expert system was described[112] for the study of small sets of compounds which were of special interest. This system in turn was based upon a statistics-based diagrammatic scheme for classifying the entire database of known structures of binary, ternary and quaternary compounds and their tendencies to form compounds in binary and ternary alloy systems. When applied to quasicrystals, ferroelectrics and superconductors it revealed diagrammatic regularities which aided the recognition of phenomenological trends and guided computerized search strategies for the identification of promising new materials. A rapid and exhaustive approach was later applied[113] to the materials classification problem in which every feature, such as a chemical property, every pair of features, every triple of features, and so on, was to be tested against every classifier among a group of so-called support vector machines. That is, the featured were treated as points in space which were plotted so that those of different types were divided by a maximal clear gap. New examples were then mapped into the same space and were predicted to belong to a given category on the basis of to which side of the gap they fell. This generalized the original classification scheme to higher dimensions and re-affirmed that the Mendeleev number was the best guiding parameter, especially when combined with the valence-electron number. The prediction of structures via data-mining of quantum results typically involved precise quantum-mechanical computations performed on a limited set of candidate structures, or the use of empirical rules gleaned from extensive experimental data. This approach offered limited predictive power, and it was suggested[114,115] that the use of heuristic rule-extraction methods should be applied to a large library of *ab initio* derived information and thus be developed into a better tool for structure-prediction. Promising outcomes were reported[116] in systematic searches for new ferroelectric, ultra-hard, ductile or high-dielectric/permittivity materials.

A further algorithmic approach was developed[117] for the prediction of crystal structures by combining data-mining with quantum mechanics. Machine-learning methods identified the physical principles which governed structure-choice, and quantum mechanical principles then added the final accuracy. An informatics-based structure-suggestion model proposed probable ground-state structures which fitted a specific prediction and thus guided a large-scale analysis of intermetallics. *Ab initio* structure calculations were performed using density functional theory in the generalized gradient approximation, and experimental results indicated that the relationship between structures at various compositions in a given alloy was essentially the result of similar atomic interactions between the constituent atoms. As a concrete success, structure-maps were found[118] for $A^I_4A^{II}_6(BO_4)_6X_2$ apatite compounds by data-mining. The problem as always was to identify the key crystallographic parameters that could act as coordinates for the structure-map. Their selection from a large set of potential choices was achieved here via a linear data-dimensionality reduction method. A multivariate data-set of known apatites having the above form was used, and each member compound was represented as a 29-dimensional vector in which the components were discrete scalar descriptors of electronic and structural attributes. A structure-map which was defined by two distortion angles, α (the rotation-angle of A^{II}-A^{II}-A^{II} triangular units) and ψ (the angle which the A^I-O1 bond makes with the c-axis when $z = 0$ for the A^I site), was shown to classify such apatites on the basis of the occupancy of the A, B and X sites. The classification was finalized by means of a clustering analysis.

It was later shown[119] for the first time that, by using methods based upon information-entropy, it was possible to explore quantitatively the relative influences of a wide array of electronic and chemical bonding parameters in governing the structural stability of intermetallic compounds. By using the inorganic AB_2 compound database as a template, the evolution of crystal-chemistry design rules based upon an information-theoretic partitioning scheme for high-dimensional data was examined. The value of this data-mining approach to identifying optimum chemical and structural design rules was demonstrated by showing that, when combined with first-principles calculations, statistical inference techniques could markedly accelerate the prediction of crystal structures. A novel method[120] for evaluating the similarity of material compositions involved calculations, based upon data-mined ion-substitution information, of the probability that two ions could replace each other within a given structure-type. The method was checked by attempting to predict the structures of oxides in the Inorganic Crystal Structure Database. The correct structure-type was found, within 5 guesses, in 75% of the tests. Success was especially high for quaternary oxides, where the correct structure-type was identified on the first guess, in 65% of the tests. Studies of this type were made ever easier by the increasing power of computers and by evolving database design[121]. An example of the use of high-throughput methods was provided[122] by zirconia and its then poorly-understood doping behavior. Hundreds of very large-scale density functional theory defect calculations were carried out for doped cubic zirconia systems and these clarified the dilute-limit stability of almost all industrially-relevant cations on the zirconia lattice. In order to determine what factors ultimately guaranteed dopant stability in zirconia, so-called clustering-ranking modeling was used to identify just what reliable criteria were contained within the extensive property databases. The present method took account of experimental and computational results and highlighted the features of dopant oxides which best predicted their stability when dissolved in zirconia.

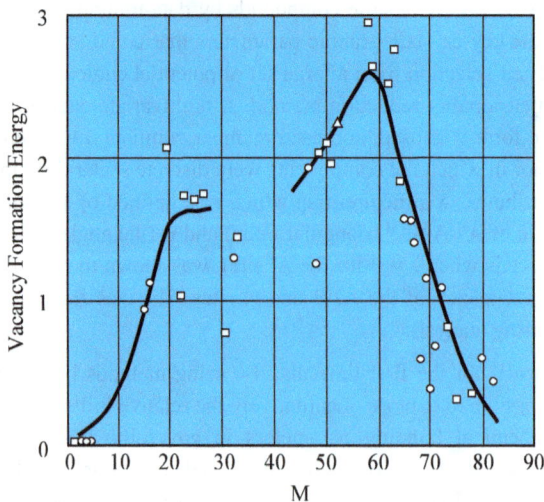

Figure 27. Face-centered cubic vacancy formation energy as a function of the Mendeleev number. Circles: face-centered cubic elements, squares: hexagonal close-packed elements

A similar high-throughput first-principles search was made[123] for new phases in binary alloys which were based upon platinum-group metals combined with transition-metals. Stable new compounds were predicted to exist in 28 binary systems where no such compounds had been shown experimentally to exist (figure 26). Dozens of other unreported compounds were predicted to exist in other systems. Also identified were stable structures at compositions which had been examined without providing structural data, suggesting that some reported compounds were unstable at low temperatures. Improved structure-maps were plotted for the binary alloys of platinum-group metals, and were deemed to be much more predictive than those based only upon empirical results. Half-Heusler semiconductors having very low thermal conductivities were found[124] by means of high-throughput computational pre-screening. Nanostructuring has further encouraged interest in thermoelectric energy conversion, with nanograined materials having higher figures-of-merit than those of bulk equivalents and thus exhibiting increased conversion efficiency. First-principles modeling of figures-of-merit was used to examine further 75 nanograined compounds which had been chosen, via electronic and thermodynamic sifting, from 79057 known half-Heusler compounds[125]. In many cases, the resultant figures-of-merit were far above those which could be achieved in nanograined IV- and III-V-type semiconductors; with some 15% of them being even better at high temperatures.

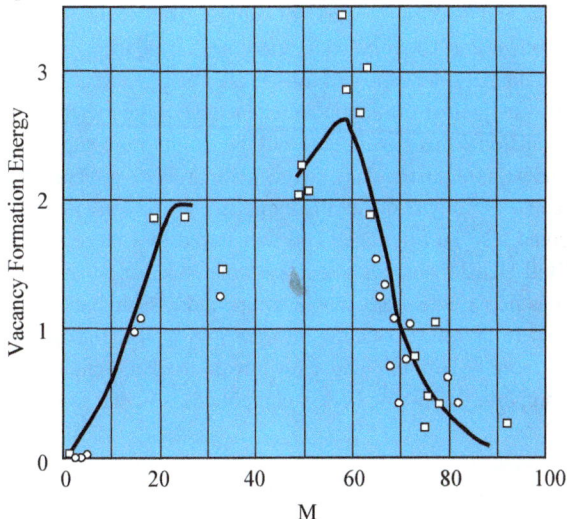

Figure 28. Hexagonal close-packed vacancy formation energy as a function of the Mendeleev number. Circles: face-centered cubic elements, squares: hexagonal close-packed elements

Materials Research Forum LLC
https://doi.org/10.21741/9781644902011

Machine-learning identified simple rules for predicting, merely on the basis of its composition, whether a nanograined half-Heusler compound will be a good thermoelectric material. Half-Heusler compounds, with a valence-electron count of 8 or 18, are promising high-temperature thermoelectric materials. This is due to their very good electrical and mechanical properties, combined with high-temperature thermal stability. With the help of first-principles calculations, great progress has been made in half-Heusler thermoelectric materials. Efficient methods for estimating the bulk conductivity for a large number of compounds were based upon machine-learning algorithms and automated *ab* initio calculations and involved scanning some 79000 half-Heusler database records. This identified 450 mechanically-stable ordered semiconductors, having conductivities which ranged over more than 2 orders of magnitude, with the conductivity being lowest for compounds where elements which were located in equivalent positions had large radii. Further screening for thermodynamic stability then reduced the list of candidates to 75. Three materials were finally identified which had conductivities of less than 5W/mK. In a similar high-throughput search for full-Heusler compounds, a machine-learning model was trained[126] to identify Heusler compounds. This model performed very well, and reliably found Heusler compounds among more than 400000 candidates; with a false-positive rate of 0.01. When applied to the screening of candidates of AB_2C-type, it predicted the existence of 12 new Heusler gallides, MRu_2Ga and RuM_2Ga, where M ranged from titanium to cobalt. In further work[127] it was shown that convolutional neural networks, using structural and chemical information from the periodic table and full-Heusler compound data, could provide lattice parameters and enthalpies of formation simultaneously. The mean prediction errors were within density functional theory accuracy and the results were superior to those obtained using only Mendeleev numbers. This suggested that the 2-dimensional structure of the periodic table could be fully and usefully exploited by the convolutional neural network. The learning could also be fine-tuned and transferred to a more informed prediction of the stability of full-Heusler compounds. It was noted that tungsten-containing compounds were identified as being potentially stable compounds. Basic band-structure calculations have been used[128] to investigate atomic disorder in n-type MNiSb compounds, where M is titanium, zirconium or hafnium, and show how to use band-engineering to improve the thermoelectric properties of p-type FeRSb compounds, where R is vanadium or niobium.

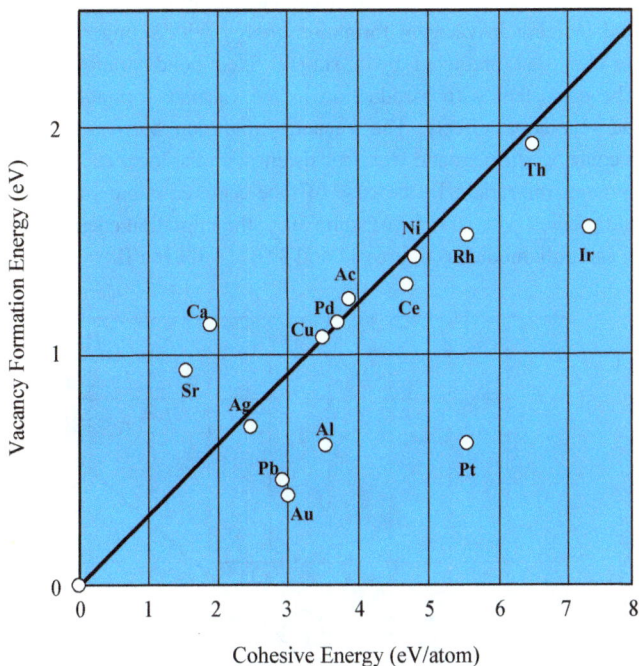

Figure 29. Face-centered cubic vacancy formation energy as a function of the cohesive energy.

With a view to predicting properties as well as structures, elemental vacancy diffusion information from a database was subjected[129] to high-throughput first-principles calculations for face-centered and hexagonal close-packed structures. The formation and migration energies for vacancies in all relevant stable pure elements were determined by performing *ab initio* calculations. In the case of hexagonal migration, both basal-plane and z-direction nearest-neighbor vacancy hops were assumed and energy barriers were calculated for 49 elements in the face-centered cubic structure and 44 elements in the hexagonal structure. These data were then plotted against various properties in a search for notable correlations, and clear trends when they were ordered using the Mendeleev numbers (figures 27 and 28). When vacancy-formation energies were plotted against cohesive energies, the trends were linear (figures 29 and 30). The slopes were 0.317 and

0.323 for the face-centered and hexagonal close-packed structures, respectively. There was an expected increase in vacancy formation energy with stronger bonding, but the slope was less than that predicted by a simple fixed bond strength model. It was considered to be consistent with a reduction in the vacancy formation energy due to many-body and relaxation effects. The vacancy migration barriers increased almost linearly with increasing stiffness and were consistent with the local expansion which was necessitated by atom migration. In the case of face-centered cubic structures, a simple expression could predict, with an error of some 30%, the migration energy in terms of the lattice constant and bulk modulus (figures 31 and 32): $H = 0.016a^3B$.

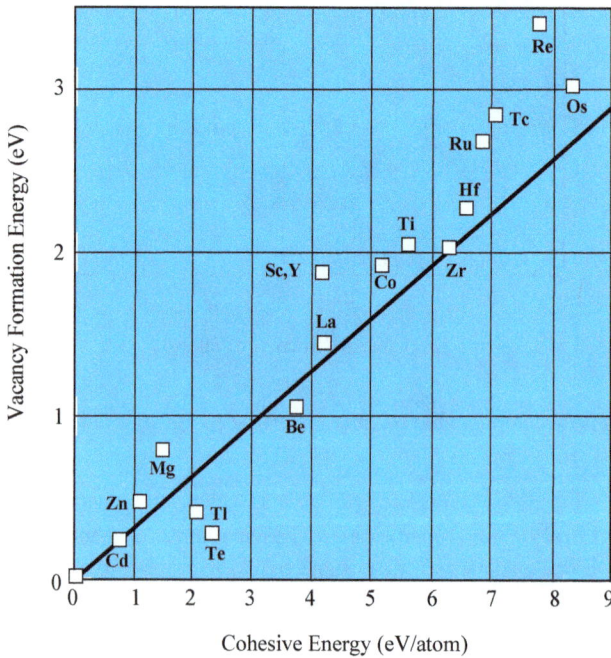

Figure 30. Hexagonal close-packed vacancy formation energy as a function of the cohesive energy.

The magnitude of the structure-prediction problem at the time was spelled out[130] by noting that, upon considering just a sub-set of 13026 ternary intermetallics, which exhibited 1391 different structure-types, 667 of them offered just a single representative. This raised the question of what subtle factors made each of them so unique. It was also noted that only 5109 ternary compounds were known among 85320 theoretically possible ternary intermetallics. On the other hand a 3-dimensional structure-map, based upon experimental data for compounds which comprised *sp*-block elements and transition metals, predicted[131] the correct crystal structure with a probability of 86% with a confidence level greater than 98% that the correct crystal structure was one of 3 predicted choices. The coordinates used were the number of valence-electrons, the atomic volume and electronegativity. When compared with density functional theory calculations for 1:1 *sp*-d-valent compounds, the 3-parameter model offered comparable success. Empirical rules still played a role. It was noted[132] that, among ternary systems involving a lanthanide, a transition metal and a p-block element, only the $HoCoGa_5$-type structure was associated with unconventional superconductivity. The reason for this was explored by preparing $ScFeGa_5$, $ScCoGa_5$ and $ScNiGa_5$; which shared the $HoCoGa_5$ structure. Their electronic structures were determined by using density functional theory and Hückel calculations. The observed valence-electron count range of $HoCoGa_5$-type was explained in terms of the so-called 18-n empirical rule; counting n = 6 for lanthanide atoms and n = 2 for transition-metal sites. Density functional theory chemical-pressure analysis of $ScNiGa_5$ revealed that the pressure was negative within the gallium sub-lattice as it expanded to house the scandium and transition-metal atoms. This explained why $HoCoGa_5$-type gallides were observed only for small lanthanide and transition-metal atoms.

The techniques of partial least-squares discriminant analysis and support vector machine analysis were used[133] to develop a structure-predictor for AB compounds, predicated on the classification of 706 such compounds – found in the Pearson Crystal Data and ASM Alloy Phase Diagram databases - which exhibited one of the 7 commonest structure-types: CsCl, NaCl, ZnS, CuAu, TlI, β-FeB, NiAs. Among the 56 original criteria, 31 were selected via forward selection and backward elimination; the quality of the model being judged by determining the cluster-resolution at each step. The partial least-squares discriminant analysis yielded a sensitivity of 96.5%, a specificity of 66.0% and an accuracy of 77.1% for the validation-set data. The support vector machine method yielded a sensitivity of 94.2%, a specificity of 92.7% and an accuracy of 93.2%. The atomic radius, electronegativity and valence-electron count were again confirmed as being pivotal variables. Both of the analytical techniques furnished quantitative predictions of unknown compound structures. The novel compound, RhCd, was predicted by partial least-squares discriminant analysis to have a CsCl-type structure, with an 0.669 probability. The support vector machine technique offered the same prediction, but with a

Materials Research Forum LLC
https://doi.org/10.21741/9781644902011

0.918 probability. Subsequent compound-preparation proved the predictions to be correct. It was naturally concluded that the support vector machine method was better. An artificial intelligence tool was also described[134] which learned to spot trends in the valence- and conduction-band characteristics of the more than 2500 compounds listed in the Materials Project database. A database of band-gaps in inorganic substances was integrated[135] with other information which was available on the properties of inorganic substances and provided a basis for finding correlations between the band-gap width and other parameters of thermoelectric materials.

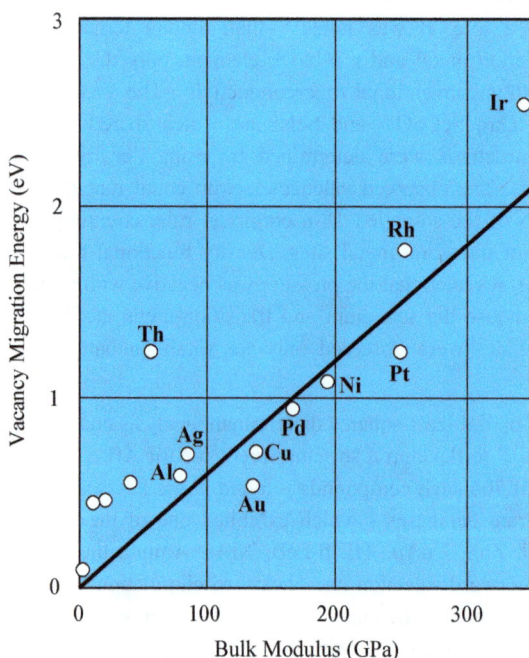

Figure 31. Face-centered cubic vacancy migration energy as a function of the bulk modulus.

A 2-dimensional structure-map was proposed[136] for the prediction of transition-metal Laves phases in terms of the atomic radius ratio and d-electron total count. The structure-map (figure 33) revealed that a radius-ratio greater than 1.116 was required for the formation of 16-coordinated Frank-Kasper polyhedra in transition-metals Laves phases. A d-electron total count of between 5 and 10 seemed to place empirical limits on the tendency to form unsaturated covalent bonds in the transition-metals systems.

In a search for $L1_2$ ternary phases which might produce high-strength superalloys, 2224 metallic systems were scanned[137] for precipitate-hardening phases of the form: $X_3[A_{0.5},B_{0.5}]$, where X was nickel, cobalt or iron and [A,B] could be lithium, beryllium, magnesium, aluminium, silicon, calcium, technetium, scandium, titanium, vanadium, chromium, manganese, iron, cobalt, nickel, copper, zinc, gallium, strontium, yttrium, zirconium, niobium, molybdenum, ruthenium, rhodium, palladium, silver, cadmium,

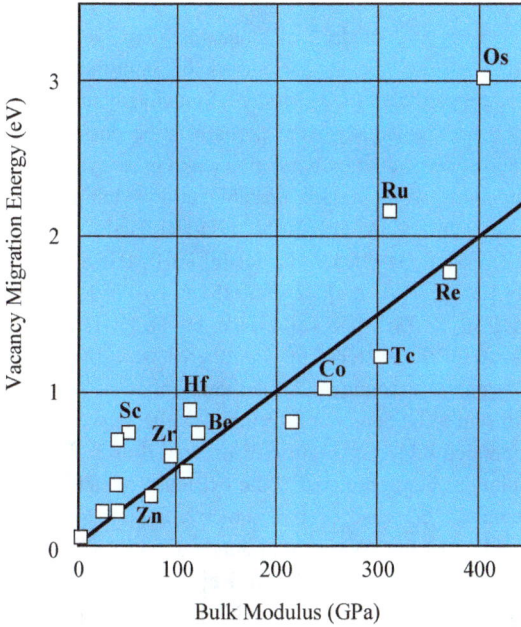

Figure 32. Hexagonal close-packed vacancy migration energy as a function of the bulk modulus.

indium, tin, antimony, hafnium, tantalum, tungsten, rhenium, osmium, iridium, platinum, gold, mercury or thallium. It was found that 102 systems exhibited a lower decomposition energy and a lower formation enthalpy than those of the $Co_3(Al,W)$ superalloy. They entered into a stable 2-phase equilibrium with the host matrix, $X_3[A_xB_{1-x}]$, where x ranged from 0 to 1, and the lattice mismatch was less than 5%. Of the 102 systems, 37 had not been explored but elimination of the obviously too-harmful (beryllium, thallium, technetium) and too-expensive (gold, iridium, silver, osmium, platinum) allowed the choice to be pared down to just 6 promising candidates. There was a time when half of the 28 known rhodium phase diagrams with transition metals were reported to be phase-separating or incomplete. Extensive high-throughput first-principles calculations[138] then predicted the existence of stable ordered structures in 9 of the 14 binary systems, and also predicted some unreported compounds to exist in the known compound-forming systems.

A similarly extensive study was made[139] of binary hafnium systems with alkali metals, alkaline earths and transition metals by means of high-throughput first-principles calculations. These again predicted the existence of previously unknown compounds in 6 binary systems which had previously been thought to be phase-separating. They also predicted the existence of some unknown compounds in other systems and suggested that some reported compounds were in fact unstable at low temperatures. Results were reported for the systems: Ag-Hf, Al-Hf, Au-Hf, [Ba-Hf], Be-Hf, Bi-Hf, [Ca-Hf], Cd-Hf, Co-Hf, Cr-Hf, Cu-Hf, Fe-Hf, Ga-Hf, Hf-Hg, Hf-In, Hf-Ir, [Hf-K], [Hf-La], [Hf-Li], Hf-Mg, Hf-Mn, Hf-Mo, [Hf-Na], [Hf-Nb], Hf-Ni, Hf-Os, Hf-Pb, Hf-Pd, Hf-Pt, Hf-Re, Hf-Rh, Hf-Ru, Hf-Sc, Hf-Sn, [Hf-Sr], [Hf-Ta], Hf-Tc, Hf-Ti, Hf-Tl, [Hf-V], Hf-W, [Hf-Y], Hf-Zn and [Hf-Zr], where the systems in parentheses were those for which *ab initio* methods predicted no stable compounds. At that time only 6 of 28 rhenium systems with transition metal were said to be compound-forming, while 15 were reported to be phase-separating and 7 contained high-temperature disordered θ or ξ phases. High-throughput first-principles calculations here predicted[140] the existence of stable ordered structures in 20 of the 28 systems. In the case of the known compound-forming systems, these predictions reproduced all of the known compounds together with some unknown ones. Among the investigated systems: [Ag-Re], [Au-Re], [Cd-Re], Co-Re, [Cr-Re], [Cu-Re], Fe-Re, Hf-Re, [Hg-Re], Ir-Re, Mn-Re, Mo-Re, Nb-Re, Ni-Re, Os-Re, Pd-Re, Pt-Re, Re-Rh, Re-Ru, Re-Sc, Re-Ta, Re-Tc, Re-Ti, Re-V, [Re-W], Re-Y, [Re-Zn] and [Re-Zr], parentheses again indicate those for which the *ab initio* methods predicted no stable compounds. In the case of 28 binary ruthenium systems involving transition metals, high-throughput first-principles calculations predicted[141] the existence of unknown compounds in 7 of 16 systems which had previously been thought to be phase-separating, and in 2 of 3 systems which were thought to contain only a high-temperature σ-phase. Some unknown compounds were predicted to exist in 5 more systems, while some known compounds were suggested to be unstable at low temperatures. Among the systems investigated: [Ag-Ru], [Au-Ru], [Cd-Ru], [Co-Ru], [Cr-Ru], [Cu-Ru], [Fe-Ru], Hf-Ru, [Hg-Ru], Ir-Ru, Mn-Ru, Mo-Ru, Nb-Ru, [Ni-Ru], Os-Ru, [Pd-Ru], Pt-Ru, Re-Ru, Rh-Ru, Ru-Sc, Ru-Ta, Ru-Tc, Ru-Ti, Ru-V, Ru-W, Ru-Y, Ru-Zn and Ru-Zr, those in parentheses were predicted to contain no stable compounds. This method is most valuable when predicting the constitution of systems which would be technically difficult to examine experimentally. This is true for instance of the radioactive transition metal, technetium. It was noted that half of 28 systems involving technetium were reported to be phase-separating or incomplete. High-throughput first-principles calculations predicted[142] the existence of stable ordered structures in 9 of the above 14 binary systems. They also predicted the existence of new compounds in all 9 of the known compound-forming systems and in 2 of the 5 systems which were reported to contain disordered χ- or σ-phases.

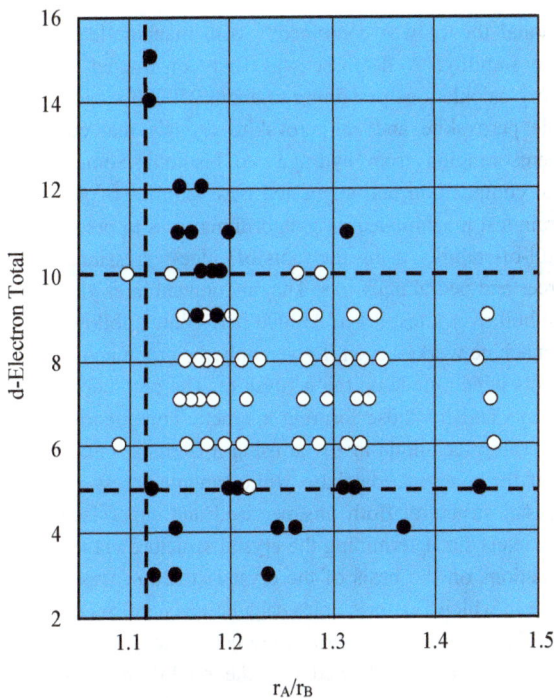

Figure 33. Criteria for predicting transition-metal Laves phases. The vertical line is at $r_A/r_B = 1.116$. Black: non-Laves, white: Laves.

So-called sub-group discovery was proposed[143] as a data-mining approach to data which were generated by density functional theory calculations. It was shown that sub-group discovery could identify physically meaningful criteria which labeled the crystal structures of 82 octet binary semiconductors as being either rock-salt or zincblende. Sub-group discovery also revealed a 2-dimensional model, involving only the atomic radii of valence s and p orbitals, which accurately classified the crystal structures of 79 out of 82 octet binary semiconductors. The sub-group discovery approach was also applied to 24400 configurations of neutral gas-phase gold clusters, containing 5 to 14 atoms, in order to distinguish general patterns among geometrical and physicochemical properties. In particular, sub-group discovery revealed that van der Waals interactions within the gold clusters were linearly related to their radius of gyration, and were weaker for planar clusters than for non-planar clusters. A parameter was also found which predicted that a local linear correlation existed between the chemical hardness and cluster isomer stability of even-numbered gold clusters.

Density functional theory was combined[144] with machine-learning in order to predict thermodynamic stability. A database was first constructed which contained density functional theory calculations pertaining to about 250000 cubic perovskite, including all of the possible perovskite and antiperovskite crystals that could be fabricated using feasible elements ranging from hydrogen to bismuth. Some of the phases were of unconventional composition and suggested new families of perovskites. This database was used to train/test machine-learning algorithms so as to predict the energy-distance to the convex hull-of-stability using the tools of ridge-regression, random forests, highly-randomized trees and neural networks. The randomized trees yielded the smallest error in distance to the hull over a test-set of 230000 perovskites, following training using 20000 samples. The machine worked even if given only the group and row, in the periodic table, of the 3 elements which made up the perovskite. The prediction accuracy was worse for first-row elements and for those forming magnetic compounds. In another approach[145], the machine-learning technique involved using a Gaussian mixture model to understand the structure of the materials database while random-forest classification was used to predict the crystal structure. Both unsupervised and supervised machine-learning then identified parameters for determining the crystal structure via the database. Prediction of atomic combinations on the basis of the crystal structure was also possible by using a trained machine in which first-principles calculations confirmed the stability of predicted materials. Machine-learning algorithms were used[146] to determine structures, given only the composition. The results showed that the random-forest technique, with Magpie features, usually out-performed other algorithms for the binary and multiclass prediction of crystal systems and space groups. A multiple-layer perceptron with atom frequency features was best for carrying out structural polymorphism prediction. A multiple-layer perceptron with atom frequency, and binary relevance with Magpie, were best for predicting crystal systems and space groups, respectively. The results also confirmed some key contributing factors to structure prediction, such as electronegativity, covalent radius and Mendeleev number.

A recent non-empirical approach to the prediction of useful material properties among all of the possible combinations of all of the available elements, combined[147] a so-called co-evolutionary method with a restructured Mendeleevian chemical space, energy filtering and Pareto optimization. This ensured that any predicted material possessed optimum properties with a high probability that it could in fact be created. It was confirmed that diamond and its polytypes are the hardest possible materials and that body-centered cubic iron exhibits the highest zero-temperature magnetization. A reliably predictive 3-dimensional structure-map was described[148] for stoichiometric binary *sp*- d-valent compounds. The coordinate parameters which were used were transferable to off-stoichiometric compounds, and retained their predictive power. They were also suitable for dealing with ternary prototypes. A 3-dimensional structure map was constructed for 129 prototypical crystal structures which were applicable to ternary compounds, and the structure was predicted correctly with a probability of 78%. The correct crystal structure was expected, with a confidence of 95%, to be among the 3 most likely crystal structures predicted by the structure-map.

A method for predicting the crystal structure of equiatomic ternary compositions was based[149] only upon the nature of the constituent elements. It was developed by using cluster-resolution feature-selection and support-vector machine classification. In the first stage using supervised machine-learning, the model was trained with the aid of 1037 individual compounds that adopted the most common ternary 1:1:1 structure types of TiNiSi, ZrNiAl, PbFCl, LiGaGe, YPtAs, UGeTe and LaPtSi. It was then checked by using a further 519 compounds. Cluster-resolution feature-selection then improved the class discrimination and indicated that 113 variables, including size, electronegativity, valence-electron number and position in the periodic table, governed the choice of structure. The prediction sensitivity, specificity, and accuracy of the final model were 97.3, 93.9 and 96.9%, respectively. It was concluded that this method was capable of predicting the crystal structure, given only its composition. The powers of cluster-resolution feature-selection and support-vector machine classification were further demonstrated by separating the crystal structures of polymorphs in order to consider polymorphism in the TiNiSi- and ZrNiAl-type structures. Upon analyzing 19 compositions which were known to adopt both structures, the machine-learning model correctly identified, with a confidence level of better than 70%, the low-temperature polymorph; given its high-temperature form. The machine-learning model also showed that some compositions could not be clearly differentiated and lay within a so-called confused region where the confidence level was between 30 and 70%. This implied that both polymorphs might well be observable within in a single sample under certain experimental conditions. Later synthesis and examination of a single TiFeP sample which exhibited both TiNiSi- and ZrNiAl-type structures, even after 3 months of annealing, confirmed the existence of the confused region of structural uncertainty which had been predicted by the machine-learning technique.

Data-mining machine-learning techniques, together with the non-equilibrium Green's function method and density functional theory, were used[150] to study the electronic transport properties of an organic-inorganic hybrid perovskite. The band structures, derived from a first-principles analysis showed that the ferroelectric and antiferroelectric dipole configurations had very little effect upon the energy band-gap. The gradient-boosting regression tree model was found to be the most effective one among the various algorithms available for the rapid prediction of electron transmission coefficients. The general trend has indeed been for structure and property prediction to progress from heuristics, or structure-maps based upon *ad hoc* parameters, to the use of parameters which have been identified by applying machine-learning techniques to large databases containing experimental and/or theoretical calculations[151].

The elastic properties of inorganic materials were predicted[152] by using 3 feature-selection methods and 3 machine-learning algorithms: linear regression, ridge regression and support-vector regression. The best feature subset was first selected in order to predict the elastic properties of inorganic compounds from among a large number of multi-scale feature sets. The most effective model that combined both feature-selection and machine-learning was then found for the prediction of the elastic properties of materials. The performances of various combinations of feature-selection methods and machine-learning models were finally compared by analyzing the feature subset which was obtained using various feature-selection methods. Experimental results subsequently showed that the use of filtering plus support-vector regression resulted in the best predictive performance. The machine-learning model exerted a greater influence upon the prediction results than did the feature-selection method. The feature subset which was selected by the feature-selection method included mostly characteristics such as the melting point, crystal structure and Mendeleev number.

In earlier work[153], aspects of the crystal structure of an optical material had been used to construct an information-based model for the determination of the tendency of a given AB composition to exhibit multiple crystal systems. This involved the exploitation of both supervised methods, such as support-vector machines, and unsupervised methods, such as disorder-reduction and principal-component classification. Examination of the relative contributions made by materials-chemistry descriptors revealed that the Mendeleev number played an important role, but had to be weighed against the effects of cationic/anionic radius ratios and electronegativity differences between the constituents of the unit cell. It was also necessary to consider the role which was played by dynamic charges in the classification of octet AB-type binary compounds into four-fold structures such as zincblende/wurtzite, and six-fold structures such as rock-salt. It was shown[154] that the difference in dynamic charges of the four-fold and six-fold coordinated structures, together with Harrison's polarity, were excellent parameters for classifying the coordinations of 82 *sp*-bonded binary octet compounds. A support-vector machine was used to estimate the classification accuracy and variance of the present model, and a decision-boundary was found under supervised conditions. Similar methods were used[155] to classify the crystal structures of the octet subset of AB solids and to predict their melting-points. Some possible problems were found with the use of machine-learning methods when using relatively small data-sets. On the other hand, an important new material parameter, the so-called excess Born effective charge was identified which markedly increased the accuracy of predicted classifications. A new scale was also proposed for the degree of ionicity and covalency in such solids. It was then possible to classify the structures of a set of 75 octet solids into those that have ionic or covalent bonding. By using long-established indices it was possible to achieve an average classification success rate of 92%. The use of just one traditional index, plus the new excess Born effective charge of the A-atom, yielded an average success rate of 97%. There were also relatively large variations however around these averages which depended upon how the machine-learning methods were used. The usual standard deviation was unsuitable for quantifying the degree of confidence placed on these averages. On the other hand, it could be stated, with 95% confidence, that the traditional classification parameters gave an accuracy of between 89 and 95%, and that the accuracy of the newly proposed pair of parameters was between 96 and 99%. When predicting the melting-points, the size of the data-set was 46 and the root-mean-square error of the model was 11% of the mean melting-point of the data. When the accuracy of this predicted error was itself measured, the estimated fitting error itself had a root-mean-square error of 50%. The classification and regression predictions could clearly vary to a marked extent, depending upon how machine-learning methods were applied to limited data-sets.

An interpretative machine-learning model has been used[156] to examine structure-property relationships and predict solid solubility in binary alloy systems. This involved the use of a dataset which contained some 1843 binary alloys and the associated experimentally determined values of solid solubility. A function was defined which represented the relationship between the individual descriptor and the solid solubility, and a deep neural network was used to integrate those multiple functions. The resultant model could correctly predict the solid solubility value more accurately than could previous machine-learning models, and it was additionally possible to analyze the effect of the structure of the material upon the property in question.

New compounds were sought[157], which offered an improvement in certain chosen properties, while performing the minimum possible number of experiments or calculations. This was done by applying optimum learning concepts and methods to 3 material data-sets: a set of experimental data on more than 100 shape-memory alloys, a data-set of 223 density functional theory calculations of M_2AX phases and a computational data-set covering 704 piezoelectric compounds. It was found that the Maximin and Centroid design strategies were the most efficient. Although the datasets varied in size and type, the Maximin algorithm was superior for all of the data sets, especially when the accuracy of a machine-learning model fit was not high.

In recent years materials informatics[158], which is the application of data science to problems in materials science and engineering, has emerged as a powerful tool for materials discovery and design. This relatively new field was already having a significant impact upon the interpretation of data for a variety of materials systems, including those used in thermoelectrics, ferroelectrics, battery anodes and cathodes, hydrogen storage materials, polymer dielectrics, etc. Its practitioners employ the methods of multivariate statistics and machine learning in conjunction with standard computational tools (e.g., density-functional theory) to, for example, visualize and dimensionally reduce large data sets, identify patterns in hyperspectral data, parse microstructural images of polycrystals, characterize vortex structures in ferroelectrics, design batteries and, in general, establish correlations to extract important physical information and infer structure-property-processing relationships.

Zinc-based alloys are among the most commonly used light alloys. In order to improve knowledge of the ternary systems which are formed by zinc plus two lanthanides, the isothermal section at 400C of the Er-Pr-Zn system was studied[159]. Some selected samples at the Er:Pr ratio = 1:1 were investigated by differential thermal analysis in order to draw the vertical section of the system. The experimental techniques used were X-ray powder diffraction and scanning electron microscopy, coupled with electron probe microanalysis and differential thermal analysis. No ternary compounds were found in the system. With regard to the intermetallic phases, the $(Er,Pr)Zn$, $(Er,Pr)Zn_2$, $(Er,Pr)Zn_3$, $(Er,Pr)_{13}Zn_{58}$ and $(Er,Pr)_2Zn_{17}$ solid solutions formed within the full field of compositions, while the $ErZn_{12}$, Pr_3Zn_{11} and $PrZn_{11}$ compounds partially dissolve the third element. The metastable intermetallic binary phases, $ErZn_5$, $PrZn_5$ and Er_6Zn_{23}, were not found as either ternary solid solutions nor as binary compounds in the ternary Er-Pr-Zn system at 400C.

The fundamental principles underlying the arrangement of elements into solid compounds with an enormous variety of crystal structures are still largely unknown[160]. A recent study presented a general overview of the structural types which appear in an important subset of the solid compounds, i.e., binary and ternary compounds of the 6A column oxides, sulfides and selenides. It included an analysis of these compounds, including the prevalence of various structure types, their symmetry properties, compositions, stoichiometries and unit-cell sizes. It was found that these compound families include preferred stoichiometries and structure types that may reflect both their specific chemistry as well as a research bias in the available empirical data. Identification of non-overlapping gaps and missing stoichiometries in these structure populations may be used as a guide to the search for new materials.

As part of a thermodynamic study of binary Fe-RE system, critical evaluations and optimizations of all available phase diagrams and thermodynamic data for Fe-RE alloys, where RE was gadolinium, terbium, dysprosium, holmium, erbium, thulium, lutetium or yttrium, were conducted to obtain reliable thermodynamic functions of all the phases in the systems[161]. In the thermodynamic modeling of these rare-earth systems, systematic variations in the phase diagrams and thermodynamic properties such as the enthalpy of mixing in the liquid state and enthalpy of formation of solid compounds with the atomic number of lanthanide series were observed. These systematic trends were incorporated into the optimization of the Fe-RE system in order to resolve inconsistencies among the available experimental data and to estimate unknown thermodynamic properties. The systematic trends in thermodynamic properties of solid and liquid phases and phase diagram of the entire Fe-RE systems were summarized.

Forming a four-component compound from the first 103 elements of the periodic table results in more than 10^{12} combinations[162]. Such a space of materials was intractable to high-throughput experimentation or first-principles computation. A framework was introduced in order to address this problem and quantify how many materials can exist. Principles of valency and electronegativity were used to filter chemically implausible compositions, which reduced the inorganic quaternary space to 10^{10} combinations. It was demonstrated that estimates of band-gaps and absolute electron energies can be made simply on the basis of the chemical composition and thus applied to the search for new semiconducting materials to, for example, enable the photo-electrochemical splitting of water. It was shown that it was applicable to the prediction of crystal structure by analogy with known compounds, including exploration of the phase space for ternary

Materials Research Forum LLC
https://doi.org/10.21741/9781644902011

combinations that form a perovskite lattice. Computer screening reproduced known perovskite materials and predicted the feasibility of thousands of others. Given the simplicity of the approach, large-scale searches could be performed using a single workstation.

First-principles calculations have become a powerful tool for exploring the physical and chemical properties of materials[163]. They provide fundamental information for seeking and designing advanced materials. They have been successfully applied to the prediction of new materials such as cubic BN, superconductors, electrodes, etc., and also used to investigate aspects such as mechanical properties, phase transformations, diffusion, etc. Titanium-based alloys have long found application as a class of high-temperature structural materials, due to their low density, high specific strength and good high-temperature creep resistance. They have been used as new-generation biomedical materials having a relatively low modulus and non-toxicity. Attention was here focused upon the application of first-principles theory to developing new titanium-based alloys.

The International Temperature Scale of 1990 standardized practical temperature measurements on the basis of the reproducibility of the solidification phase-transition temperature of high-purity metals[164]. The effect of impurities can be the greatest barrier to establishing the international equivalence of temperature metrology, and so it is important to have detailed knowledge of the distribution coefficients for all potential impurities. In particular, impurities having a high distribution coefficient cause problems when applying corrections, and so it is helpful to identify those impurities which are likely to create problems. By plotting published measured distribution coefficients of impurities in the metals used as references in the international temperature scale, as a function of a quantum mechanics-based scale which separated binary alloy crystal structures, an apparent resonance peak was found. This allowed a simple equation to be used to determine whether a given impurity is likely to cause errors in any correction process.

An empirical relationship between band-gap and orbital electronegativities, which was based upon Zunger's orbital radii, was identified by using a simplified bond orbital model[165]. A bonding parameter for each A-B bond among 35 elements could be constructed by using this empirical relationship. The bonding parameter could predict compound formation between *sp*-bonding atoms, to within about 90% accuracy. The error in judgement when using the bonding parameter was within about 20kJ/g-atom of the heat of formation for most of the unpredictable compounds.

The moments of the electronic density-of-states provide a robust and transparent means for the characterization of crystal structures[166]. Using *d*-valent canonical tight-binding, the moments of the crystal structures of topologically close-packed phases were calculated using density-functional theory. The moments were then used to establish a measure for the difference between two crystal structures and to characterize volume changes and internal relaxations. The second moment provides access to volume variations of the unit cell and of the atomic coordination polyhedra. Higher moments reveal changes in the longer-ranged coordination shells due to internal relaxations. Normalization of the higher moments leads to constant (A15,C15) or very similar (χ, C14, C36, μ, and σ) higher moments of the density-functional theory relaxed topologically close-packed phases across the 4d and 5d transition-metal series. The identification and analysis of internal relaxations was demonstrated for atomic-size differences in the V-Ta system and for various magnetic orderings in the C14-Fe_2Nb Laves phase.

A first-principles study was made[167] of a large set of simple highly symmetrical ordered Fe-compounds (with superlattices of body-centered cubic structure) in order to analyze the role of magnetism in the phase stability of these compounds. The results confirmed that ferromagnetism and compound stability could be related. That is, the highest magnetic moments were found for the least stable compounds. It was also shown that compounds exhibit qualitatively different behaviors with regard to stability, according to whether their nature is ferromagnetic or non-magnetic.

Alloys of $ThMn_{12}$-type structure, with the general formula $Nd_{1-x}Zr_xFe_{10}Si_2$ where x ranged from 0.4 to 0.8, were prepared[168]. X-ray diffraction and Mössbauer spectroscopy measurements indicated the formation of the tetragonal $ThMn_{12}$-type structure upon substitution of neodymium by zirconium, without further annealing. The lattice parameters, a and c, and the unit-cell volume obeyed Vegard's law; the metallic radius of the zirconium (1.602Å) substitution being smaller than that of the neodymium (1.821Å) metallic radius. The Curie temperature decreased linearly with zirconium substitution, from 301C for x = 0 to 285C for x = 0.8. The saturation magnetization and magnetic anisotropy fields were also measured as a function of zirconium substitution, with the room temperature values ranging from 12.5kG and 30kOe for x = 0.4 to 11.2kG and 26kOe for x = 0.8. The success of this synthesis contributed to the search for novel magnetic phases which are low in, or free from, strategically critical raw materials, and promise a possible new generation of permanent magnets.

Density-functional-theory based calculations were made[169] of the vacancy-formation energies in metals by using the revised Tao-Perdew-Staroverov-Scuseria (revTPSS) functional, which is a self-consistent semilocal meta-generalized gradient approximation functional. The object was to determine whether the improved accuracy of surface energies for revTPSS, as compared with local and generalized gradient approximation functional, also led to improved vacancy formation energies; since vacancies could be viewed as being internal surfaces. In addition to lattice constants, cohesive energies and bulk moduli were predicted by revTPSS. By comparing the vacancy formation energies and bulk properties, the performance of revTPSS was assessed in competition with four functionals: the local spin density approximation (LSDA), the Perdew, Burke and Ernzerhof (PBE), the Perdew-Wang-91 (PW91) and the PBE for solids (PBEsol). Using an automated computational approach, the vacancy formation energies and the macroscopic properties of 34 metal systems were calculated for these five functionals. With regard to macroscopic properties (lattice constants, cohesive energies, bulk modulus) it was found that the results for revTPSS typically lay between LDA and PBE, with a mean absolute percentage deviation of 1.1 and 12.1% from the experimental data for lattice constants and cohesive energies, respectively. It was found furthermore that revTPSS predicted higher vacancy formation energies when compared with the four other functionals. The order which was observed for the functional, with respect to the computed vacancy formation energies, was Efxc:EfrevTPSS > EfPBEsol~EfLDA > EfPBE > EfPW91. Also considered were the effects of a surface-energy error corrections that had been proposed for standard LDA and GGA functionals. This correction increased the vacancy formation energies of LDA, PBE and PW91 functionals. The revTPSS-computed vacancy formation energies were greater than the surface-energy-corrected PBE vacancy formation energies by a mean relative difference of 14.8%.

Experimental studies of compressed matter are now routinely conducted at pressures exceeding 100GPa, and occasionally at pressures greater than 1TPa[170]. The structures and properties of solids that have been so significantly compressed differ considerably from those of solids at ambient pressure (1atm), often leading to new and unexpected physical phenomena. Chemical reactivity is also substantially altered within these extreme pressure regimes. It has been shown how a synergy between theory and experiment can pave the way towards new experimental discoveries. Because chemical rules-of-thumb established at 1atm often fail to predict the structures of solids under high pressures, automated crystal structure prediction methods are increasingly useful. A few examples from the literature here exemplified just how useful theory can be when used as an aid in the interpretation of experimental data, describing innovative theoretical predictions that guide experiment and analyzing when the computational methods that are currently and routinely used in fact fail.

Phase competition between C11b, C16 and E93phases under pressure (0 to 60GPa) in $Zr_2Cu_{1-x}Ni_x$, where x was 0, 0.125, 0.25, 0.50, 0.75, 0.875 or 1, was investigated[171] by performing first-principles pseudopotential calculations. The entire picture of phase competition, mediated by both external pressure and composition, was unambiguously established by modeling while using both super-cell and virtual crystal approximation methods; thus shedding light on the pressure-dependent glass stability and crystallization behavior of related alloy systems. It was revealed that both pressure and an increasing nickel content stabilized the C16 phase with respect to the other structures. The underlying mechanism of external pressure and composition which stabilized the C16 phase was also analyzed.

Scanning electron microscopy studies involving energy-dispersive X-ray analyses of as-cast and annealed samples of Ni-Ru-Y were used[172] to produce a solidification projection, a liquidus projection surface and an isothermal section at 1200C. The ~YNi_2, ~YNi_3 and ~YRu_2 phases had wider solubilities than line compounds, and also extended furthest into the system. The binary phase extensions into the ternary system were: ~51at%Ru for ~YNi_2; ~22at%Ru for ~YNi_3; ~13at%Ru for ~YNi_5; ~7at%Ru for ~YNi; ~12at%Ni for ~YRu_2 and ~10at% for ~$Y_{44}Ru_{25}$. Ruthenium stabilized ~YNi_2, so that it solidified at a higher temperature in the ternary than in the Ni-Y binary system. A ternary phase was confirmed at $Y_{51}Ru_{15}Ni_{34}$ (at%), which formed via a peritectic reaction. The ~Y_3Ru and ~Y_3Ni phases were isomorphous and formed a continuous solid solution. Heat-treatment at 1200C yielded the phases: (Ru), ~YRu_2, ~YNi_2, ~YNi_3, ~YNi_4, ~YNi_5 and (Y).

The isothermal section at 400C of the Dy-Gd-Zn system was studied[173], and the vertical sections at the Dy/Gd ratio = 1 was investigated. No ternary compounds were found in the system. With regard to the intermetallic phases, the $(Dy,Gd)Zn$, $(Dy,Gd)Zn_2$, $(Dy,Gd)Zn_3$, $(Dy,Gd)_3Zn_{11}$, $(Dy,Gd)_{13}Zn_{58}$, $(Dy,Gd)_2Zn_{17}$ and $(Dy,Gd)Zn_{12}$ solid solutions formed within the full range of compositions, while the Gd_3Zn_{22} compound did not exhibit any dysprosium dissolution.

In advanced materials parameters such as temperature, pressure, structure, composition and disorder determine their properties[174]. Band-structure calculations have been used to identify the electronic, structural and magnetic properties of the layered ternary compound: AMn_2X_2 (A = barium, calcium or yttrium and X = tin, germanium or silicon). The calculations were performed by using the scalar-relativistic full potential linearized augmented plane wave method. The equilibrium lattice parameters, atomic positions within the unit cell, interatomic distances, band structure, density-of-states and spin magnetic moment in the unit cell were taken into account. The density-of-states indicated a typical conductive electronic structure for all of the systems. Metal-excess behavior was observed due to the high concentration of Mn-3d electrons just down to the Fermi level ε_f. The magnetic behavior introduced new features with regard to the magnetic moments for such ternary structures. The magnetic moment of the studied compounds jumped to a high value as the unit-cell volume increased to above 180Å^3 or when the Mn-X bond-length became greater than 2.5Å.

A well-defined notion of chemical compound space is essential for taking effective control of properties via variations of elemental composition and atomic configurations[175]. An atomistic first-principles perspective on chemical compound space was introduced here. The chemical compound space was first considered in terms of variational nuclear charges in the context of conceptual density functional and molecular grand-canonical ensemble theory. Then the notion of compound pairs, related to each other via interpolations involving fractional nuclear charges in the electronic Hamiltonian was reconsidered. Taylor expansions in chemical compound space, property non-linearity, improved predictions using reference compound pairs and the ability to linearize the chemical compound space were addressed. Attention was finally turned to the machine-learning of analytical structure property relationships in chemical compound space. These relationships corresponded to inferred, rather than derived through variational principles, solutions of the electronic Schrödinger equation.

The widely accepted intuition that the important properties of solids are determined by a few key variables underpins many theoretical methods[176]. Although this reductionist paradigm is applicable to many physical problems, its utility can be limited because an intuition for identifying the key variables often does not exist or is difficult to develop. Machine-learning algorithms (genetic programming, neural networks, Bayesian methods, etc.) attempt to eliminate the *a priori* need for such intuition but often do so with increased computational exertion and worktime. A then recently developed technique in the field of signal processing, compressive sensing, provided a simple, general and efficient means for finding the key descriptive variables. Compressive sensing was indeed a powerful basis for model-building; it was shown that its models were more physical and more accurately predictive than then-current state-of-the-art approaches and could be constructed at a fraction of the existing computational cost and effort.

An electronic quantity, the correlation strength, was defined as a necessary step for understanding the properties and trends in strongly correlated electronic materials[177]. As a test case, this was applied to the various phases of elemental plutonium. Within the GW approximation a so-called universal-scaling relationship was found, wherein the f-electron bandwidth reduction due to correlation effects was shown to depend only upon the local density approximation bandwidth and was otherwise independent of the crystal structure and lattice constant.

Titanium alloys can exhibit three distinct crystal structures: α, β and ω. For various applications alloying elements can be used[178] to stabilize a desired phase. Extensive data exist for determining the thermodynamic equilibrium phase and, typically, phase coexistence. The normal state of commercial alloys was however a quenched solid solution. While alloy designers had well-established rules-of-thumb, any rigorous theory of non-equilibrium single-phase crystal stability was less well-established. A theory was developed, for predicting which phase a particular alloy will adopt, as a function of minor element concentrations. Two different methods were used, based upon density functional theory with pseudopotentials and plane waves, with either explicit atoms or the virtual crystal approximation. The former was highly reliable, while the latter made a number of drastic assumptions that typically led to poor results. The agreement between the methods was nevertheless surprisingly good, showing that the approximations which were made in the virtual crystal approximation were not important in determining the phase stability and elastic properties. This permitted a generalization which showed that the single-phase stability can be related linearly to the number of d-electrons, independent of the actual alloying elements or the details of their atomistic-level arrangement. This leads to a quantitative measure of β-stabilization for each alloying transition-metal.

It had been argued that the heat of formation of intermetallic compounds was mostly concentrated in the nearest-neighbor unlike-atom pair-bonds, and that the positive term in Miedema's equation was associated[179] with charge transfer on the bond so as to maintain electroneutrality. Taking examples of some well-populated crystal-structure types such as $MgCu_2$, $AsNa_3$, $AuCu_3$, $MoSi_2$ and $SiCr_3$, the effect of such charge-transfer upon the crystal structures adopted by intermetallic compounds was examined. It was shown that the correlation between the observed size-changes of atoms upon alloying and their electronegativity differences supported the idea of charge transfer between atoms. It was argued that the electronegativity and valence differences need to be of the required magnitude and direction to alter, through charge transfer, the elemental radius ratios R_A/R_B to the internal radius ratios r_A/r_B allowed by the structure types. Since the size change of atoms upon alloying was closely related to how different R_A/R_B was, from the ideal radius ratio for a structure type, the lattice parameters of intermetallic compounds could be predicted with excellent accuracy, knowing R_A/R_B.

In order to perform atomistic simulations of steel, it is necessary to have a detailed understanding of the complex interatomic interactions in transition metals and their alloys[180]. The tight-binding approximation provides a computationally efficient, yet accurate, method for investigating such interactions. In this work an orthogonal tight-binding model for iron, manganese and chromium, with the explicit inclusion of magnetism, was parameterized from *ab initio* density-functional calculations.

The effect of the alloying elements tantalum, molybdenum, tungsten, chromium, rhenium, ruthenium, cobalt and iridium upon the elastic properties of both γ-nickel and γ'-Ni_3Al was studied[181] by using first-principles methods. Results for the lattice properties, elastic moduli and the ductile/brittle transitions were all presented. The calculated values agreed well with existing experimental observations. The results

Materials Research Forum LLC
https://doi.org/10.21741/9781644902011

showed that all of the additions decreased the lattice misfit between γ and γ' phases. Different alloying elements were found to have differing effects upon the elastic moduli of γ-nickel. Whereas all of the alloying elements slightly increased the moduli of γ'-Ni_3Al, apart from cobalt, both of the phases became more brittle with alloying element additions; again apart from cobalt. The electronic structures of the γ' phase when alloyed with various elements were provided as examples for elucidating the various strengthening mechanisms.

It was noted that $FeGa_3$ was unusual for an intermetallic compound in that it had a semiconducting gap of the order of $0.5eV$[182]. Conventional density-functional based electronic-structure calculations in the local-density approximation gave a similar gap but it was expected that iron, a 3d transition metal, was likely to display an on-site Coulomb repulsion, U, that should be taken into account; particularly in an insulating compound with some narrow bands. First-principles local-density approximation calculations for $FeGa_3$ were analyzed and then adjusted so as to include on-site Coulomb repulsion in a mean-field way (LDA+U method) in order to show that, with a moderate U-value of the order of 2eV, iron moments were obtained in anti-aligned Fe_2 dimers (band-theory so-called singlets) with a band-gap that still coincided with the observed gap. Increasing the U-value gradually and counter-intuitively reduced the gap and finally produced an incorrect metallic state. It was suggested that more experimental effort should be made to distinguish between the so-called Fe_2-singlet and non-magnetic descriptions, and that calculations of the optical properties should be provided for comparison with data.

Under normal conditions, sodium forms a 1:1 stoichiometric compound with indium, and also with thallium, both in the double-diamond structure[183]. But sodium does not combine with aluminum at all. The question was posed as to whether NaAl could exist and, if so, under what conditions and with what structure. Instead of beginning with a purely computational and first-principles structure search, early Brillouin and higher (Jones) zone ideas of the phenomena determining structure-selection were applied. The higher zone concept, as applied to the stability of metals and intermetallic compounds, was extended to problems where density becomes a primary variable, within the second-order band structure approximation. An analysis of the range of applicability of pressure-induced Jones zone activation was presented. The simple NaAl compound served as a numerical test-bed for the application of this concept. Higher zone arguments and chemical intuition led quite naturally to the suggestion that 1:1 compound formation between sodium and aluminium should be favored under pressure, and lead specifically to a double-diamond structure. This was confirmed computationally by density functional theoretical methods within the generalized gradient approximation.

Materials Research Forum LLC
https://doi.org/10.21741/9781644902011

Activation energies for vacancy-mediated impurity diffusion in face-centered-cubic aluminum have been computed[184] *ab initio* for all technologically important alloying elements, as well as for most of the lanthanides. The so-called 5-frequency rate model was used to establish the limiting vacancy interchange process. Many elements were shown to be limited by aluminium-vacancy interchanges. For these elements it was shown that the diffusion activation energy was quite close to that for aluminium self-diffusion, and additionally the diffusion pre-exponential factor was of the same order as that for aluminium self-diffusion. The diffusion activation energy was shown to exhibit a linear relationship to the solute partial molar volume in aluminium. On the other hand transition metals were shown to deviate strongly from these generalities. Diffusion of transition-metal atoms was limited by solute-vacancy interchanges that require remarkably high activation energies. Transition-metal diffusivities in aluminium show strong trends with the number of d-valence electrons but not with partial molar volume.

A computational study showing that high pressure fundamentally alters the reactivity of the light elements lithium and beryllium[185], which were the first metals in the condensed state and immiscible under normal conditions. Four stoichiometric Li_xBe_{1-x} compounds were identified that were stable over a range of pressures, and it was found that the electronic density of states of one of them displays a remarkable step-like feature near to the bottom of the valence band, and then remains almost constant with increasing energy. These characteristics were typical of a quasi-2-dimensional electronic structure, the emergence of which in a three-dimensional environment was rather unexpected. This was attributed to large size-differences between the ionic cores of lithium and beryllium: as the density increases, the lithium cores start to overlap and thereby expel valence electrons into quasi-two-dimensional layers which are characterized by delocalized free-particle-like states in the vicinity of beryllium ions.

The thermal properties and elastic constants of Nd-Mg ordered intermetallic alloys were studied using molecular dynamics simulations[186,187]. The calculated results were in good agreement with the available experimental data and the first-principles data at various temperatures. The calculated elastic constants decrease as the temperature increases, which indicates that thermal softening occurs. Moreover the thermal volume expansion values of NdMg, $NdMg_2$, $NdMg_3$, Nd_5Mg_{41} and $NdMg_{12}$ were 7.26×10^{-5}, 7.96×10^{-5}, 7.54×10^{-5}, 8.29×10^{-5} and 8.02×10^{-5}/K, respectively, at 298K. The results indicate that among those intermediate phases, the interatomic force of NdMg was the strongest, and the melting point of the phase was the highest, which was also reflected by the bulk modulus. In addition the heat capacities of NdMg, $NdMg_2$, $NdMg_3$, Nd_5Mg_{41} and $NdMg_{12}$ were 23.46, 23.32, 23.48, 23.38 and 23.40J/molK, respectively, at 298K.

It was shown that two sharply defined physical numbers (nuclear charge, number of valence electrons) and two coarse fuzzy parameters (ranges of energies, spatial extensions of atomic orbitals) characterize the atoms of the chemical elements and determine the two-dimensional structure of the periodic system[188]. Some relevant facts concerning quantum chemistry were reviewed and it was concluded that the important factor in chemistry was the overall set of non-diffuse orbitals in low-energy average configurations of chemically bonded atoms. A decisive aspect of periodicity concerned the energy gaps between the core and valence shells. Diffuse Rydberg orbitals and minute spin-orbit splittings were important in spectroscopy, but less so with regard to chemical science and the periodic system.

A key issue in metallic glass research has been a conceptual understanding of their formation[189]. The richness of this topic arises from an interplay of various factors such as thermodynamics, kinetics, electronic structure and local geometry. A brief review of the various approaches to understanding glass-formation introduced very simple schemes for integrating the various factors, thereby creating a synergistic model for understanding glass formation.

A novel approach was used to infer the existence of structures on the basis of combinatorics and geometrical simplicity[190]. The method identifies least random structures, for which the energy was an extremum (maximum or minimum). Although the key to the generic nature of the approach was energy minimization, the extrema were found in a chemistry-independent manner.

The binary alloys of rhodium, palladium, iridium and platinum were considered[191], and it was shown that the $Rh_{1-x}Ir_x$ and $Rh_{1-x}Pt_x$ systems, which had been thought to exhibit phase separation at low temperatures, in fact exhibited miscibility over the entire concentration and temperature ranges. Low critical ordering temperatures were found, indicating that long-range order was unlikely to be observed experimentally. The results were compared with previous theoretical predictions for the other binary alloys of rhodium, palladium, iridium and platinum and with new calculations for the $Pt_{1-x}Ir_x$ and $Pd_{1-x}Ir_x$ systems. The mechanisms which decided the ordering or phase-separating behavior of all 6 binary alloys among the above quartet were investigated, particularly with regard to why, among its chemically similar members, 3 of the binary alloys (PdIr, PtIr, RhPd) exhibited phase-separation while the remaining 3 (RhIr, RhPt, PdPt) exhibited ordering.

The occurrence of interstices, their number and position in cubic crystal structures was investigated[192] from three viewpoints: physical metallurgy, crystallography and quantum-mechanical computations. Interstices were deemed to be anything but just leftover free space. They play an important role in governing the stability or alteration of crystal structures, because they are able to include electronic charge density. Together with structural and thermal vacancies they are a universal means of reacting to the influence of those physical factors that govern the formation of the observed structure. The description of the positions of interstices in a crystal structure can be accomplished in the simplest case by viewing the unit cell, by computing the Wigner-Seitz cells of the atoms in the unit cell or by computing the electronic charge density distribution. The latter two steps are very expensive however. The interaction of the so-called Fermi surface necks with the {111} planes of the first Brillouin zone of copper and the non-occurrence of a structural alteration in this case was interpreted as being a direct consequence of the charge-containing ability of the interstices.

The Ho-Al-Cu isothermal section at 500C was investigated[193] over the entire composition range, revealing the ternary phases: $Ho_3(Cu_xAl_{1-x})_{11}$ (x = 0.12 to 0.185), $Ho(Cu_xAl_{1-x})_3$ (x = 0.215 to 0.41), $Ho(Cu_xAl_{1-x})_2$ (x = 0.43 to 0.615), $Ho(Cu_xAl_{1-x})_{12}$ (x = 0.36 to 0.55), $Ho_2(Cu_xAl_{1-x})_{17}$ (x = 0.42 to 0.69), $Ho(Cu_xAl_{1-x})_5$ (x = 0.45 to 0.815), $Ho_6Cu_{15.4}Al_{7.6}$, $Ho(Cu_xAl_{1-x})_6$ (x = 0.805 to 0.83). The phase equilibria resulting from the large number of binary and ternary compounds were determined. The general characteristics of the section were considered in comparison with those of other R-Al-Cu systems.

A new binary antimonide, Ti_2Sb, was found[194] to crystallize in a distorted variant of the La_2Sb type, which contains a square planar lanthanum net with short La-La bonds. In the Ti_2Sb structure, the corresponding titanium net was deformed to squares and rhombs in order to enhance Ti-Ti bonding, as proved by single-crystal X-ray investigation combined with the real-space pair distribution function technique utilizing both X-ray and neutron powder diffraction data. Electronic structure calculations revealed a lowering of the total energy which was caused by the disorder, the major driving force being strengthened Ti-Ti interactions along the diagonal of the Ti_4 rhombs.

Laves phases constitute the largest group of intermetallic phases. Although they are long well known, there are still unsolved problems concerning the stability of the respective crystal structures[195]. The Laves phases crystallize with a cubic $MgCu_2$- or a hexagonal $MgZn_2$- or $MgNi_2$-type structure which differ only by the particular stacking of the same 4-layered structural units. It was still not possible to predict which of the structure types was the stable one for a Laves phase compound AB_2. Phase transformations from a cubic low-temperature structure to a hexagonal high-temperature structure were observed as well as stress-induced transformations from the hexagonal structure to the cubic one. In addition, deviations from the stoichiometric composition were reported to result in a change of the stable polytype in various systems. In dealing with fundamental aspects of the stability of Laves phases, some factors which are known to affect the occurrence and structure type of Laves phases were considered and it was shown that previous models and calculations were not well-suited to furnishing a general description of the stability of Laves phases.

Treatment of different alloy phenomena allowed to study[196] extended ground-state searches for α-brass and Ag-Pd detected the existence of low-temperature ground states as an expansion of the corresponding phase diagrams. The results showed that the face-centered cubic based system, the heavily segregating zinc atoms led to a grain boundary between the segregated zinc layers and the aluminium-rich Al-Zn alloy driven by the energy gain of the zinc layer by giving up the unstable face-centered cubic phase.

Two procedures were suggested[197] to determine sublimation enthalpies and bond energies between the components A and B of a binary alloy. One procedure was based on the determination of the deposition potential of the less noble metal at the alloy, another method was connected with the evaluation of characteristic underpotential deposition potentials. The interaction energy was one of the factors determining formation of intermetallic compounds. Examples were given of the expected and observed structures in electrochemical deposits.

Electron back-scattered diffraction and energy dispersive X-ray spectroscopy were performed on a plate-shaped phase[198] which formed by the reaction of tin and nickel. The phase was formed during extended thermal cycling tests on ceramic capacitors having electroplated tin end terminations. The morphology was identical to that of the phase, $NiSn_3$. The phase was shown to have the stoichiometry, $NiSn_4$, and a crystal structure isomorphous to $PdSn_4$, $PtSn_4$ and $AuSn_4$. The structure could also be described as being a higher-symmetry structure).

The zinc-rich corner of the 450C isothermal section of the Zn-Fe-Co system was experimentally determined[199]. ζ-$FeZn_{13}$ and ζ-$CoZn_{13}$ were found to form a continuous solid solution. As a result, the liquid and the ζ phases form a sub-binary system at the zinc-rich corner of the ternary system. In addition to the ternary extensions of the binary δ-$FeZn_{10}$ and δ-$CoZn_{10}$ phases, a ternary compound, designated as T-phase in the present study, has been identified. X-ray powder diffraction studies suggest that the T-phase was isomorphous to Γ'-Fe_5Zn_{21}. The limits of the solid solubility of iron and cobalt in their counter δ phases are appreciable, being 3.1 and 4.2at%, respectively.

From a phenomenological analysis of phase stabilization in the binary actinide alloy systems, U-Pu, U-Np-Pu and U-transition metals, the existence of a relatively stable electronic complex with the configuration $(5f\,nd)^v$ was deduced[200]. It must be formed by itinerant 5f and nd electron states with n = 3, 4, 5, 6 and v should be a fractional number since only band states are involved. The stability of the complexes should depend upon temperature. In α-Pa, equal numbers of 5f and 6d states in the valence electron system result: (i) in a (5f6d)* super-complex which correlates with the unique crystal structure of α-Pa and (ii) with metal properties which rather resemble those of α-thorium than that of α-uranium. The alleged properties of the complex were compared with the thermal behavior of α-proactinium and of the α-uranium upon approaching the α/β transition temperatures and proactinium should play a key role for understanding the temperature effects in the light actinide series. The literature on the fabrication and properties of proactinium metal was critically reviewed with regard to a comparison between the relevant properties of α-thorium, α-proactinium and α-uranium.

The problem of thermodynamic modeling has been widely discussed and approached from various directions[201]. From the crystallochemical point of view, two types of question are worth answering: a) the crystallographically sound thermodynamic description of a complex phase, having a structure corresponding to several sets of equipoints (or Wyckoff positions) and b) the sub-division of the atoms which pertain, within a certain space group, to a specific Wyckoff position into two or more groups of equipoints of lower multiplicity, when some ordering takes place, i.e. on passing from a structure to its derivative ordered variants. It is notable that a coherent crystallographic approach is very useful in modeling the thermodynamic properties of the phases involved. Various important structure-types (σ, μ, χ, Laves) have been analyzed and solutions found for convenient thermodynamic modeling in concert with a correct crystallographic description. It is well known, on the other hand, that alternative, complementary descriptions of the various structure types are available. In selected systematic collections, critically revised descriptions of the structure types have been provided which include, for each prototype, not only the Pearson symbol, space group, lattice parameters and occupied Wyckoff positions, but also a description of the coordination (coordination polyhedra and next-neighbor histograms). On the other hand the role of the coordination, described in terms of so-called atomic-environment types has been used for the systematic presentation of various prototypes and their possible classification and grouping into inter-related families. These aspects are relevant to the discussion of alloy-system properties and seeking general criteria for their thermodynamic modeling within the framework of the compound energy formalism. Examples of their use in the thermodynamic modeling of a selected phases were presented.

Intermediate phases in the zirconium-rich region of the Zr-Nb-Fe system have been investigated[202]. The chemical composition ranges covered by the alloys were 41 to 97at%Zr, 32 to 0.9at%Nb and 0.6 to 38at%Fe. The phases found in this region were: the solid solutions, α-zirconium and β-zirconium, the intermetallic Zr_3Fe with less than

0.2at%Nb in solution, two new ternary intermetallic compounds $(Zr+Nb)_2Fe$ '$\lambda 1$' with a cubic Ti_2Ni-type structure in the range of 2.4 to 13at%Nb and 31 to 33at%Fe and $(Fe+Nb)_2Zr$ '$\lambda 2$' indexed as hexagonal Laves phase $MgZn_2$ type (C14) with a wide range of compositions close to 35 to 37at%Zr, 12 to 31at%Nb and 32 to 53at%Fe.

It was well known that an accurate prediction of phase equilibria in multi-component systems requires a sound thermodynamic characterization[203] of the binary and ternary sub-systems involved. In turn this requires good sets of data which are precisely determined and critically assessed. Examples of semi-empirical criteria which might be useful in selection and assessment procedures were reported.

A systematic search for metal-tin systems which show a potentially high re-melting temperature in diffusion-soldered bonds was performed[204] by investigating phase formation and reaction kinetics in M-Sn systems, where M was zirconium, hafnium, niobium, tantalum or molybdenum. Phase equilibration experiments were carried out using powder samples which were prepared from pure elements, heated in evacuated silica capsules between 300 and 1000C for 1h to 200 days, and studied by X-ray diffraction analysis. Diffusion couples, M/Sn, where M was zirconium, niobium, tantalum or molybdenum were prepared from foils of pure elements and heated in a bonding furnace or in evacuated silica tubes at 300 to 700C for 6 to 95h. The rates of M-Sn reactions could be semi-quantitatively ordered in the sequence: Zr > Nb > Hf > Ta ≫ Mo (-Sn). The activation energies of the better-defined reactions. Zr-Sn, Nb-Sn and Hf-Sn were in the range of 67 to 93kJ/mol. A Ta-Sn phase diagram was constructed which had similarities to the V-Sn system. Only Zr-Sn was a potential candidate for the diffusion-soldering of high-temperature stable bonds, molybdenum was suggested to be an effective diffusion barrier against attack by liquid tin.

The alloy, Ti-46.5Al-2Cr-3Nb-0.2at%W, with a fully lamellar microstructure was subjected to creep deformation at 1073K and 270MPa[205]. Following creep-deformation, fine lamellae consisting of γ and α_2 laths and β precipitates were observed, which did not exist in the undeformed material, and these observations were explained in terms of $\alpha_2 \rightarrow \gamma$ and $\alpha_2 \rightarrow \beta$-phase transformations. In order to understand the mechanisms of the phase transformations, an overall investigation was performed by combining transmission electron microscopy, high-resolution electron microscopy, energy-dispersive spectroscopy and fast Fourier transform techniques. It was found that the two types of phase transformation were closely related, but proceeded via differing mechanisms. The $\alpha_2 \rightarrow \gamma$-phase transformation was a form of stress-induced phase transformation, while the $\alpha_2 \rightarrow \beta$-phase transformation was due to the segregation of chromium and tungsten.

A parameter-free approach based upon *ab initio* density functional calculations was shown[206] to describe the phase stability and order-disorder transformations in Pd-V substitutional alloys and intermetallic compounds with remarkable accuracy, thus permitting first-principles calculations of the complete alloy phase diagram. The investigations were based upon electronic structure and total-energy calculations for ordered compounds and disordered alloys (treated in a super-cell approximation) using gradient-corrected exchange-correlation functionals and a plane-wave-based all-electron method. All of the calculations involved a complete optimization of all structural degrees-of-freedom. The calculation of the free energies of the competing phases was based upon quite simple mean-field descriptions of long-range and short-range ordering phenomena, using concentration-dependent interchange and shell interaction parameters. In addition, the electronic structures of ordered compounds and of substitutional alloys were analyzed.

A method was proposed[207] for calculating the electronegativity of metal atoms in crystalline solids on the basis of the thermochemical and structural parameters of inorganic materials. The electronegativities thus calculated were lower than those for metal atoms in a molecular state; the difference decreasing as the oxidation state of the metal becomes higher, and vanishing for tetravalent metals.

Iron solubilities in molten Zn-Al alloys were experimentally determined at temperatures from 450 to 480C, a range which is relevant to continuous galvanizing operations[208]. The iron solubility was found to decrease slowly with increasing aluminium content in regions where ζ (FeZn$_{13}$) or δ (FeZn$_7$) was the equilibrium compound and rapidly in the region where the η (Fe$_2$Al$_{15}$Zn$_x$) phase was the equilibrium compound. Analyses of the experimental data indicated that the iron solubility was governed by the thermodynamic properties of the intermetallic compound in equilibrium with the molten Zn-Al alloy. A model was developed in order to describe the liquid surface in the Zn-rich corner of the Zn-Fe-Al system. The methodology developed here proved to be applicable to the determination of the liquidus surface in the Zn-Fe-Ni system.

The unique ability of tessellation methods to characterize atomic structures was highlighted; with the Voronoi tessellation[209], because of its freedom from constraining assumptions, being emphasized as being pre-eminent among these methods. It was demonstrated how to extract information from an atomic structure by constructing its Voronoi tessellation and analyzing the results. In so doing insight was gained into the coordination patterns of the atoms and how their positions are influenced by neighbors. As an example of such a procedure, typical Voronoi volumes were calculated for 72 atom types found in a set of 249 binary intermetallic and ionic compounds. Whereas the various atom types have broadly characteristic volumes, these are also sensitive to the other atom types with which they are combined in a given compound, i.e. whether the latter behaves as a metal, semimetal, semiconductor or insulator. Smaller variations in volume occurred moreover for different atom types within these four categories. Structural information obtained from the Voronoi tessellation could thus possibly be correlated directly with atomic properties.

The cluster units which are obtained by capping all of the faces of a tetrahedron (tetrahedral star) and a trigonal bipyramid (double tetrahedral star) were used[210] as building units in order to describe and rationalize the framework structures which are found in the intermetallic structure-types of $NaBa$, $NaZn_{13}$, Th_6Mn_{23}, $Ba_2Li_{4.21}Al_{4.79}$, $BaHg_{11}$, $BaCd_{11}$, $Cr_{23}C_6$, β-Mn, $Ba_3Li_3Ga_{4.1}$, $BaLi_4$, $CaZn_3$, $EuMg_{5.2}$, $ErZn_5$, Sr_3Mg_{13} and $Sr_9Li_{17.5}Al_{25.5}$. The electronic requirements for the optimum structural stability of these frameworks in the case of sp-bonding were investigated using the simple tight-binding Hückel model. As a result the networks exhibited a pronounced maximum of stability in the range of 2.1 to 2.6 electrons per atom, with the particular optimum values depending slightly upon the kind of basis cluster and its connectivity. Considering the sp-bonded representatives of the above-listed structure-types, the frameworks which were described by the basic clusters, tetrahedral star and double tetrahedral star, were usually formed by the divalent metals, beryllium, magnesium, zinc, cadmium and mercury or a combination of a mono- and a trivalent metal, e.g., (Li,Al), (Cu,Al) or (Ag,Al). More electropositive atoms, such as the heavier alkaline earth metals, calcium, strontium and barium, were embedded in such frameworks. Upon applying a formal electron-transfer from these atoms to the slightly more electronegative framework-forming atoms, the values obtained for the framework valence election concentration were very close to the optimum calculated ones. It was therefore argued that this family of intermetallic compounds could still be interpreted as electron compounds according to the Zintl-Klemm concept. The new representative, $BaLi_7Al_6$, which was synthesized by fusion of the elements and characterized by single-crystal X-ray diffraction methods, supported this idea. The $BaLi_7Al_6$ was isotypical to $NaZn_{13}$ (Fm3c, a = 12.9377Å, Z = 8) with the zinc sites occupied by mixed lithium and aluminium atoms.

The first measurements were made[211] of the structure factor, S(Q), and radial distribution function, G(r), of yttrium oxide in the normal and supercooled liquid states in the temperature range of 2500 to 3100K. Data were obtained using synchrotron X-ray scattering on levitated laser-heated liquid specimens. At temperatures far above the melting point, the first and second coordination shells began to merge, indicating increased ionicity in the liquid. As the temperature was lowered into the supercooled region, there was a substantial sharpening and strengthening of the first peaks in both

S(Q) and G(r). Supercooling caused a decrease in the first-shell coordination and Q values, together with an increase in the Y-O interionic distance.

The electronic structures of the then-recently discovered $AuCo_{2(1-u)}Sn_4$, (where u is between 0.167 and 0.180) ternary phases and the $AuNi_2Sn_4$ crystal structures which belong to the NiAs family, were investigated[212] by using local-density-functional theory and the linear muffin-tin orbital method. It was found that the stability of $AuCo_{2(1-u)}Sn_4$ was governed by strong *d-sp* bonding occurring in the low-symmetry (monoclinic) Ni_3Sn_4-type structure, whereas $AuNi_2Sn_4$ was stabilized essentially by nearly free electron *sp*-bonding in the rhombohedral Fe_3S_4 *s*-type (smythite mineral) structure. A systematic analysis of the relative total energy across the *3d* transition metal (T) row for the ternary AuT_2Sn_4 series showed that the Fe_3S_4 *s*-type structure tended to be stable when the valence electron number per T-atom (N) was less than 3.0, or greater than -9.0, whereas the Ni_3Sn_4-type structure was more stable when N was between 3.0 and 9.0. The calculations were in agreement with the fact that vacancies have to be introduced into $AuCo_{2(1-u)}Sn_4$ in order to stabilize this phase. The spin-polarized electronic structure calculations were in excellent agreement with magnetic susceptibility measurements for both phases.

Absolute measurements of the L-edge core-level binding energies were performed[213] for nickel-aluminium compounds by using electron-energy-loss spectroscopy. The core-level shifts were found to compare favorably with *ab initio* calculations of the valence band shifts. The measured electron-energy-loss spectroscopy oscillator strengths and core-level shifts were used to test the local charge neutrality approximation to self-consistency in extended Hückel tight-binding calculations. Within the local charge neutrality approximation, the tight-binding calculations could provide estimates of both the core-level shifts and, with the use of the force theorem, the alloy heats of formation for the Ni-Al intermetallics.

The phase equilibria and crystal structures of various intermediate phases were presented[214]. Results were obtained by experimental investigations of the Y-Al and Sm-Al systems. The enthalpies of formation of YAl_2 and $SmAl_2$ were re-measured, resulting in values of -50.5kJ/(mol at) and -55.0kJ/(mol at) at room temperature; in good agreement with literature data. Phase-equilibria investigations of the Sm-Al system were carried out and the results of thermal analysis, micrographic examination, microprobe and X-ray diffraction analyses were examined within the framework of the general behavior of rare-earth alloys. The experimental Sm-Al phase diagram, combined with previous literature data, was compared with the results of thermodynamic optimization. Various intermetallic compounds were found to exist: Sm_2Al (peritectic decomposition, at 860C) SmAl (peritectic decomposition at 960C), $SmAl_2$ (melting at 1480C), $SmAl_3$

(peritectoidal decomposition at 1130C), Sm_3Al_{11} (melting at about 1380C). Eutectic equilibria were discovered or confirmed to exist at 20at%Al and 760C, 75at%Al and 1340C, 97.0at%Al and 635C. An eutectoidal equilibrium was found to occur at 10at%Al and 700C.

The general properties of magnesium alloys with one or two different rare-earth metals were briefly summarized[215]. The regular trends which are observed in intermediate phase formation were used to predict the characteristics of the room-temperature isothermal section of the Gd-Y-Mg system. The partial experimental section at 500C in the composition range of about 50 to 100at%Mg was found to be in fairly good agreement with that predicted. The determination was carried out using X-ray powder diffraction, optical and scanning electron microscopy and electron probe microanalysis. The phases which were identified were a continuous solid solution of cP2-CsCl type along the line $Gd_{1-x}Y_xMg$ and, along the line $Gd_{1-x}Y_xMg_2$, two Laves-type solid solution fields (cF24-$MgCu_2$ type, $GdMg_2$ based for $0 \leq x < 0.18$ and hP12-$MgZn_2$ type, YMg_2 based for about $0.26 < x \leq 1$). Extension into the ternary field was determined moreover for magnesium and the $GdMg_3$, $GdMg_5$ and Y_5Mg_{24}-based phases.

Idealized geometrical models for the χ-gallium and β-gallium structures were based[216] upon two differently-distorted corrugated 3^6 nets. An investigation was made of the structural stability of two sets of 2-dimensional and 3-dimensional model structures consisting of these corrugated nets as a function of the net puckering. The simple tight-binding Hückel model was used together with the structural energy difference theorem and the advanced full-potential linear muffin-tin orbital method within the framework of local-density functional theory. Both methods revealed the existence of an optimum puckering-angle that corresponded well to the situation in the experimental χ-gallium structure. The geometrical model of the χ-gallium structure followed the building principle of terminally coordinated deltahedral clusters, extended to two-dimensional structures. The chemical bonding in χ-gallium was interpreted in terms of multicentre bonding within the corrugated 3^6 nets and two-electron two-centre bonds connecting those nets.

Popular techniques for analyzing the spatial and energy characteristics of chemical bonding in solids based[217] upon Hückel theory, the Hartree-Fock method and electron density functional theory were reviewed. Methods for calculating the total energies and dependent characteristics (cohesion energies, formation energies, partial pressures, etc.), the moments of the densities of states, bond occupations, and the pair potentials of electron density and localization function were considered. Examples of using these calculations for high-melting and laminated compounds were provided.

The Voronoi method for constructing polyhedra about all atoms in a given crystal structure was extended so as to provide[218] a general, quantitative and concise, means of distinguishing crystal structures from one another. Although applicable to all crystal structural types, the methodology was developed here by reference to binary non-molecular inorganic systems, A_mB_n, corresponding mainly to ionic crystals and intermetallic compounds. Consideration was given to the geometrical forms of the Voronoi polyhedra, with AB_2 compounds being singled out for illustrative purposes. Two methods were proposed for summarizing the structural geometry. One was the use of representative points for elements of Voronoi polyhedra (i.e. corners, edges and faces) and relating these to space-group symmetry. The other was the calculation of the ratio of the volumes of Voronoi polyhedra, V_A/V_B. Topological structural properties (i.e. numbers of $A \cdots A$, $A \cdots B$ and $B \cdots B$ pairwise interactions), by comparison, could be quantified by the indices I_{AA}, I_{AB}, I_{BA} and I_{BB}, whose meanings were defined and elucidated. By adopting statistically disordered structures as reference points, indices I'_{AA}, I'_{AB}, I'_{BA} and I'_{BB} could be defined, leading to the generation of two-dimensional topological structure diagrams with I'_{AA} and I'_{BB} as axes. A geometrical-topological diagrams could alternatively be generated, with V_A/V_B and $I'_{AA}-I'_{BB}$ as axes.

Investigations of the ternary systems M/Li/Al of the heavier alkaline earth metals, calcium, strontium and barium yielded[219] three new intermetallic compounds which exhibited interesting structural relationships: $Ca_6Li_xAl_{23-x}$ (x ~ 11, $Fm\overline{3}m$, a = 13.430Å, Z = 4) crystallizes in the Th_6Mn_{23} structure with all sites of the majority component being occupied by lithium and aluminium. Meanwhile $Sr_9Li_{7+x}Al_{36-x}$ (x ~ 10.5, $P6_3/mmc$, a = 9.638Å, c = 27.586Å, Z = 2) was a relative of the Sr_3Mg_{13} structure, and $Ba_2Li_{3+x}Al_{6-x}$ (x ~ 1.2, $R\overline{3}m$, a = 9.946Å, c = 26.992Å, Z = 9) was an hexagonal relative of the Th_6Mn_{23} type. The sub-structures formed by lithium and aluminium atoms are described in terms of two simple basis clusters which offer structural flexibility in their connectivity. The relative stability of the sub-structures was investigated within the simple tight-binding Hückel model. Calculated energy difference curves as a function of the valence electron concentration for these networks went through a pronounced stability maximum in the region of 2.5 to 2.7 electrons per network atom. This result fitted the composition of the intermetallic compounds when applying the Zintl-Klemm concept. In addition, Mulliken population analyses of extended Hückel calculations were used to investigate the site occupancies.

Another method was based[220] upon the Hückel or tight-binding theory in which an explicit pairwise repulsion was added to the generally attractive forces of the partially-filled valence electron bands. An expansion which was based upon the power-moments of the electronic density of states was used, and the structural energy difference theorem was introduced. The repulsive energy was found to vary linearly as the second power moment of the electronic density of states. These results were then used to show why there was such a diversity of structures in the solid state. The elemental structures of the main group could be rationalized by using these methods. It was the third and fourth power moments, which corresponded in part to triangles and squares of the bonded atoms, which accounted for most of the elemental structures of the main-group elements of the periodic table. This then served as an entry into further rationalizations of transitions for noble metal alloy, binary and ternary telluride and selenide, and other intermetallic structures. A cohesive picture of both covalent and metallic bonding could thus be presented, illustrating the importance of atomic orbitals and their overlap-integrals.

It was shown[221] how a modified tight-binding method, which was based upon second-moment scaling of the electronic density of states, can be used to rationalize both crystal structure and electron counting rules. The counting rules which were considered were the Hückel 4N + 2 rule, the VSEPR model, the octet rule, Wade's rule for electron-deficient clusters and the Hume-Rothery electron concentration rules for pure metals and alloys. Covalent and metallic structures were considered which were based upon elements from group 6 to group 17 of the periodic table. It was shown finally how both the metallic and covalent bonding found in such systems can be placed into a single cohesive bonding picture, with the method of moments proving useful.

Some further developments in the application of the moments method to structural problems have been described[222]. They included a discussion of the origin of Pauling's second rule, the heats of mixing of some transition metal oxides, surface problems, insights into the complexity of solids and means for capturing the electronic aspects of disorder. Lee's conceptual framework for understanding the structures of the elements and alloys, and the electronic underpinnings of the Hume-Rothery rules, were reviewed.

The relative stability of three-dimensional nets formed by gallium or aluminium in intermetallic phases was investigated[223] within the simple tight-binding Hückel model. In order to compare structures having differing coordination numbers, the bond-lengths of the nets were changed according to the structural energy difference theorem. The structural stability as a function of the electron concentration per atom agreed well with the composition of the corresponding intermetallic compounds. The third and fourth moments of the density of states qualitatively explained the energy-difference curves but were insufficient to reproduce all of the trends in stability.

It was demonstrated[224] that the seemingly complex stacking sequence changes which are observed in topologically close-packed phases can be interpreted in terms of the ground-state phases derived from a simple one-dimensional stacking model involving only a few interplanar interactions. It was also shown that the energetics which are involved in creating various possible stacking defects in such phases can be assessed by using this model.

First-principles local density approximation calculations were performed[225] on aluminium-magnesium intermetallic compounds. Experimental studies of this system have shown that complex topologically close-packed structures with large unit cells are favored over simpler planar structures. It was shown that this was due to the opening of a quasi-gap in the density-of-states at the Fermi level; the results for the total energies confirmed that this lowered the energy.

It was first shown how simple topological restrictions, set by stoichiometry and the local coordination number, have a large influence[226] on the generation of locally symmetrical arrangements of atoms. By introducing an orbital picture which provides an electronic underpinning for the valence sum rules of Pauling and Brown, it was shown how the local structure was determined by the electronic configuration. The use of the method of moments finally permitted conclusions to be drawn concerning the identity of nearest-neighbor atoms and linkages of various types as a function of the electron count. By studying the behavior of the fourth moment as a function of an order parameter, it was shown how ordered arrangements of atoms and bonds are electronically favored.

The phase diagram and thermodynamic properties of ZrN alloys were modeled[227] by presupposing the existence of mononitrides with their metal sub-lattices having face-centered cubic, hexagonal close-packed and body-centered cubic structures. Each phase in the system was assumed to be composed of a filled metal sub-lattice and a partially filled non-metal sub-lattice. The lattice stability parameters of the hexagonal close-packed and body-centered cubic mononitrides were estimated from the salient features of the phase diagram. The thermodynamic properties of the alloys were predicted with reasonable accuracy without any particular effort being made to fit the model parameters to existing experimental data.

Second-moment scaled Hückel theory, with the inclusion of ionic terms, was used[228] to rationalize or predict the superstructures found in elemental selenium, LnQ_2 (Ln = La, Ce; Q = S, Se), $Ln_{10}Se_{19}$ (Ln = La, Ce, Pr, Nd, Sm) and $RbDy_3Se_8$. All of these structures contained distortions of square lattices of chalcogen atoms. In the case of $Ln_{10}Se_{19}$ and $RbDy_3Se_8$ these lattice distortions were coupled to ordered defects in the chalcogen square lattice. On the basis of the energy calculations eight ground state $RbDy_3Se_8$ superstructure patterns among the 5×10^5 possible alternatives were proposed. The results of the calculations were finally explained by using HOMO-LUMO and Madelung-energy arguments.

It is well known that, within the family of rare-earth elements (especially the so-called trivalent ones), several properties change according to well-recognized and systematic patterns[229]. A general appreciation of the constitutional properties (thermodynamic properties, crystal structures, phase diagrams) of the alloys formed by the various rare earths permits a number of empirical regularities to be deduced. This behavior can constitute a prediction scheme and a reliability criterion for the evaluation of the data concerning series of various rare earth alloys with the same element. Examples of the application of this behavior to phase-diagram assessment and prediction were considered. Particular attention was paid to the binary rare earth alloys which are formed from magnesium and aluminium, and predicted versions of the Tb-Mg and Tb-Al phase diagrams were constructed.

It was shown[230] that second-moment-scaled Hückel theory can be used to account for the bond-length variations which are found in elemental gallium, borohydride, transition-metal carbonyl and hydrocarbon structures. Among the systems investigated were gallium, $Os_5(CO)_{16}$, $Ir_4(CO)_{12}$ and $[Re_4(CO)_{16}]^{2-}$.

Octahedral-type coordination by halogens in the liquid state has been reported[231] for a number of trivalent metal ions on the basis of diffraction and Raman scattering experiments on their molten trihalides and from the Raman scattering spectroscopy of liquid mixtures of trihalides with alkali halides. The available data on bond-lengths and Raman frequencies were analyzed by treating an isolated $(MX_6)^{3-}$ species within a model which adopted charged-soft-sphere interionic potentials, supplemented by accounting for ionic polarization. The trivalent metal ions that were considered were M = lanthanum, cerium, praseodymium, neodymium, samarium, gadolinium, dysprosium and yttrium for X = chlorine and M = aluminium for X = fluorine. The main result of the analysis was the prediction of trends in the soft-sphere repulsive parameters for the trivalent metal ions, leading to estimates of all of the vibrational frequencies and the binding energies of such octahedral species.

It was demonstrated[232] that a tight-binding Hamiltonian, to which a pairwise additive repulsive potential has been added, can qualitatively and semi-quantitatively account for the elemental structures of gallium, boron, zinc, cadmium and manganese. It was shown that these structures result from the interplay of the number of valence electrons, the overlap of atomic orbitals and geometrical features such as the number of triangles of bonded atoms and the angles between bonds.

The three elastic stiffness constants of various nickel-based intermetallic Ni_3X, where X was manganese, iron, aluminium, gallium, germanium or silicon with the $L1_2$ structure have been determined[233] at room temperature by using the rectangular parallelepiped resonance method. The elastic anisotropy factor, A, of the compounds was generally large, ranging from 2.5 to 3.3; except for Ni_3Ge, with A = 1.71. The large anisotropy was believed to be at least partly responsible for the positive temperature dependence of the flow stress which is observed for these compounds. The elastic constants exhibit some correlation with the lattice constants. The minor element, X, can be regarded as exerting a so-called chemical pressure which causes expansion of the nickel lattice, thus resulting in a decrease in the elastic constants.

Calculations[234] were made of the liquid pair structure of YCl_3 as a prototype for strongly ionic melts of lanthanide metal trichlorides. The calculations adopted an ionic model involving a pair potential of the Busing type, with parameters for the Y^{3+} ion adjusted to the Y-Cl bond length in the YCl_3 crystal and to the breathing-mode frequency of the $(YCl_6)^{3-}$ octahedral unit. The model was solved within the hypernetted chain approximation and the results were tested against neutron diffraction data on the Faber-Ziman structure factor and the total pair correlation function of molten YCl_3. The calculated partial structure factors and pair correlation functions provide an insight into the origin and nature of the short-range and intermediate-range order in the melt.

A tight-binding bond model was used[235] to make a quantitative study of electronic interactions in nickel-titanium compounds. The model was not an *ab initio* calculation but it required only data on the elemental properties of pure metal constituents that are readily available in the literature. Combined with an extended cluster Bethe lattice method for face-centered cubic and body-centered cubic based solid solutions and the B2 phase, and the recursion method for $NiTi_2$ and Ni_3Ti stoichiometric compounds, this approach permitted the energy of formation of these alloys to be calculated and the chemical trends in the ordering effects to be analyzed.

The continuous non-metal metal transition, which occurs upon dissolving metals in molten salts, can be shifted[236] to higher metal concentrations, in solutions which involve polyvalent metal ions, by the possibility of these metals going into lower oxidation states. A microscopic model was evaluated for these processes, for the specific case of solutions of sodium metal in molten cryolite ($AlF_3 \bullet 3NaF$). The structure of the ionic melt could be understood and calculated at the microscopic level in terms of a dominant six-fold coordinated trivalent state of the aluminium ion (the $(AlF_6)^{3-}$ complex) with some admixture of a four-fold coordinated trivalent state; the $(AlF_4)^-$ complex. The sodium metal was assumed to enter the ionic liquid in the form of monovalent ions and electrons. The calculations demonstrated how these added components break up the structure of the ionic melt so as to yield localization by the formation of aluminium ions in reduced-valence states, and provide order-of-magnitude estimates for the free-energy changes that are involved in the processes. It was found in particular that, with increasing metal concentration, the equilibrium between $(AlF_6)^{3-}$ and $(AlF_4)^-$ shifted in favor of the latter, while Al^{3+} ions were released into the melt; binding the available electrons to form Al^{2+} and Al^{3+} ions. The latter species eventually become the most stable ones and also destabilize the $(AlF_4)^-$ complex.

Ample experimental evidence is available in the literature on the geometry and the stability of local coordination for polyvalent metal ions in molten mixtures of their halides with alkali halides[237]. Schemes for classifying this evidence have long been considered. Dissociation of tetrahedral halocomplexes in good ionic systems can be viewed as being a classical Mott problem of bound-state stability in a conducting matrix. Structural coordinates can more generally be constructed from the properties of the component elements in order to separate-out systems, having long-lived fourfold or sixfold coordination, and to distinguish between them.

Empirical many-body potentials were derived[238] for noble-metal alloy systems within the framework of the Finnis-Sinclair model which was based upon a second-moment approximation to the tight-binding density of states for transition metals. The most important extension of the model was a simple incorporation of interspecies interactions which involved fitting the alloying energies. The importance of properly accounting for the local atomic relaxations when constructing the potentials was emphasized. The observed principal features of the phase diagrams of the alloys are well-reproduced by using this scheme. Reasonable concentration dependences of the alloy lattice parameter and elastic constants were moreover obtained. This led to the suggestion that the fine details of the electronic structure may be less important in determining atomic structures than are more obvious parameters, such as atomic sizes and binding energies.

The phase equilibria of aluminium-containing ternary systems were reviewed[239]. The ternary systems were of the form of aluminium - refractory metal (Ti, Zr, Hf, V, Nb, Ta, Mo) - (V, Cr, Mn, Fe, Co, Ni, Cu, Zn). A trend was noted in the crystal structure of the Al_3X-type compounds as a function of the location of the constituent elements within the periodic table.

Evidence was presented[240] for believing that tin atoms, dissolved in Ni_3Fe, are located mainly on iron sites. It was shown that the hyperfine field at a tin nucleus was determined by the moments in its first- and second-neighbor shells, and that its magnitude was consistent with that found in other magnetic materials.

Materials Research Forum LLC
https://doi.org/10.21741/9781644902011

A study of the electronic structure of NiSi and NiAl was carried out by using electron spectroscopy and theoretical calculations[241]. Experimental results, resulting from X-ray photo-emission and bremsstrahlung isochromal spectroscopy, were interpreted by means of density-of-states and matrix element calculations for the compounds in their crystal structure. This offered a detailed picture of the electronic states over the entire bonding-antibonding region below and above the Fermi level. Cluster calculations which were based upon a molecular orbital approach for NiAl provided further insight into the bonding characteristics of the states of various symmetries. It was found that the simple d-p bonding scheme that was then used for silicides, in which the metal d-states were referred to as 'non-bonding', required revision.

Precise results of first-principles electronic-structure calculations of the structural and electronic properties of the binary compound, LiAl, were presented[242] for the body-centered-cubic based B32 and B2 crystal structures, the face-centered-cubic based L1$_0$ and the simple-cubic based B1 structures as obtained using the full-potential linearized augmented-plane-wave method. Particular care was taken to ensure that the convergence of the total energy, as a function of the inherent numerical parameters, yielded as precise an absolute total energy as possible. The results corroborated the findings of previous studies concerning the electronic bonding in LiAl, as well as the existence of a high-pressure phase transition from B32 to B2 that was found at 140kbar. Band structures, densities of states, total energies, and charge densities were presented, with good agreement being found with experimental data concerning the equilibrium properties.

One model yielded compound compositions[243] which were simple rational numbers and was expected to be useful for work on intermetallic phase diagrams. The occurrence of compounds in 10 binary systems (Si-Hf, Co-Y, Hg-Sr, Rh-Gd, Au-Sm, Hg-Ca, Pt-In, Au-In, Fe-Al, and Pd-Ti) was compared with the results of the calculations and it was concluded that a considerable number of compounds which occurred in these systems were justified by the proposed hard-spheres packing condition. The dependence of the efficiency of the method upon the radius-ratio and upon the components of the compound was highlighted.

A priori atomic parameters (valence orbital nodal radii) were used[244] to classify the stability of tetrahedral halocomplexes which are formed by polyvalent s-p metals in molten mixtures of their halides with alkali halides. The existing classification scheme for liquid mixtures paralleled previous structural classifications for pure components in solid and molecular phases and was also essentially consistent with a stability criterion that was based upon a Mott-type picture of a complex as being a bound state in a conducting medium.

Materials Research Forum LLC
https://doi.org/10.21741/9781644902011

The liquidus curves of 25 simple-eutectic binary alloys were constructed[245] by using the semi-empirical theory of heats of formation which was developed by Miedema. The predictions of this theory are generally quite correct. In those cases where discrepancies existed, it was possible to improve the results by retaining the formal expression of the heat of formation as proposed by Miedema, but nevertheless modifying the prescribed values of the heats of solution.

Changes in both the heat of formation of transition-metal silicides and the core-level shifts of various transition metals with silicidation were interpreted[246] in terms of the chemical trends in silicide electronic structures. It was shown that the heat of formation and the core-level shifts were dominated by transition-metal d-band occupancy which well-characterized silicide electronic structures. It was also found that the d-band occupancy was closely correlated with Schottky-barrier heights of silicide-silicon interfaces. This suggested that silicide-silicon Schottky-barrier heights were influenced not only by interface properties but also by bulk silicide electronic properties.

Structural plots were presented[247] for AB_2 molecules and solids with sp-bonding, which were based upon parameters that were derived from valence-electron orbital radii. It was shown that the same schemes that permitted the classification of crystals having different structures, were also able to distinguish molecular shapes as being linear or bent. This picture was consistent with the existence of a critical number, $N_c = 16$, of valence electrons, in agreement with the Mulliken-Walsh rule, and accounted for the many exceptions to the latter rule in cases where $N \leq 16$. New results were found for AB_3 molecules.

The equilibrium structures of sp-bonded tetramers were classified[248] on the basis of valence-electron orbital radii. Valence electron orbital radii plots reflected the transition from planar to non-planar geometries, and accounted for trends in the bond-angles, thus confirming the previously successful application to AB_2 molecules.

A simple universal model for the structural properties of sp-bonded semiconductors and insulators was presented[249,250]. This model elucidated the physical mechanisms which govern the chemical trends and predicted semi-quantitatively the stable crystal structures, bond-lengths, bulk moduli, transition pressures of structural phase transformations, long-wavelength transverse optical phonons and band structures for binary non-metals in the rock-salt, cesium chloride and zinc-blende phases. The theory explained the puzzlingly strong cation and weak anion dependence of the observed structural transition pressures. It predicted, as a function of pressure, a universal sequence of structural phase transformations among the cubic phases of binary solids. A marked softening of the transverse optical phonons across the pressure-induced phase transition, from the zincblende to the rock-salt structure in II-VI compounds, was also predicted. The physical origin of this softening was shown to be closely related to ferro-electricity. It was shown that the chemical trends in the structural properties of semiconductors and insulators are governed by a counterbalance of attractive and repulsive short-range interactions, whereas long-range interactions play only a minor role, in contrast to the classical point-charge models of ionic crystals. The theory was based upon the semi-empirical tight-binding method and included charge-transfer and non-orthogonality effects. Only properties of the neutral atoms were used as input data for a given crystal. The total energy was explicitly minimized as a function of volume in order to find the static and dynamic equilibrium crystal properties.

It was shown[251] that *a priori* atomic parameters (valence-orbital radii) permit the classification of molecular shapes for a large number of *sp*-bonded AB_2 compounds with the number of valence electrons being up to 16, in parallel with classifications of crystal structures in the solid phase. Triatomic A_3 clusters with the number of valence electrons being up to 15 were structurally classified using the same scheme.

Acoustic and optical deformation potentials in cubic IV-VI compounds were calculated[252] by diagonalizing the *p*-model Hamiltonian, including finite lattice distortions at the L- and T-points of the Brillouin zone. A very pronounced and even non-monotonic dependence was found for some deformation potentials upon the lattice deformation and, most surprisingly, $\Xi^2(L) < 0$ for the optical deformation potentials. This could explain the wide variance of experimental data for various samples and experimental conditions.

Data-Mining and Intermetallic Property-Prediction   Materials Research Forum LLC
Materials Research Foundations **128** (2022)   https://doi.org/10.21741/9781644902011

Keywords

perovskite, 58, 60, 64
Pettifor, 3, 4, 5, 6, 7, 9, 10, 11, 12, 14, 15, 16, 17, 20, 38, 43, 94, 95, 96, 97, 98, 110
phase diagram, 16, 17, 18, 19, 37, 56, 63, 74, 76, 77, 79, 83, 84, 86, 87
phase-separating, 18, 20, 39, 42, 56, 72
polymorphism, 59
pseudobinary systems, 10
puckering-angle, 80

quaternary compounds, 46

radius-ratio, 54, 87
rhombic dodecahedron, 38
rock-salt, 17, 58, 61, 88
rule-of-thumb, 4

Schottky-barrier heights, 88
s-electrons, 16, 20
semiconductors, 17, 49, 50, 58, 88
Shannon-entropy, 43
shape-memory, 16, 62
size-mismatch, 18
s-orbital, 11, 34
sp-bonded, 6, 20, 30, 61, 78, 88, 89
sp-valent, 20, 21
structure-mapping, 12, 23, 36
structure-type, 36, 48, 53, 75, 78

superalloys, 22, 55
superconductors, 45, 46, 64
superhardness, 44
s-valent, 20, 21
synergy, 66

Tao-Perdew-Staroverov-Scuseri, 65
telluric screw, 2
topologically close-packed phase, 22, 64, 82
toughness, 44
transition-metals, 49, 55

valence electrons, 6, 30, 31, 35, 41, 71, 72, 84, 88, 89
valence-electron number, 16, 46, 59
validation-set, 53
vector machine, 46, 53, 59, 60
Voronoi tessellation, 77

Watson-Bennett, 28
Wigner-Seitz, 73
Wyckoff, 75
Wyckoff positions, 75

zincblende, 17, 58, 61
Zintl, 4, 31, 78, 81
Zintl-Klemm, 78, 81
Zunger, 64, 96

About the Author

Dr. Fisher has wide knowledge and experience of the fields of engineering, metallurgy and solid-state physics, beginning with work at Rolls-Royce Aero Engines on turbine-blade research, related to the Concord supersonic passenger-aircraft project, which led to a BSc degree (1971) from the University of Wales. This was followed by theoretical and experimental work on the directional solidification of eutectic alloys having the ultimate aim of developing composite turbine blades. This work led to a doctoral degree (1978) from the Swiss Federal Institute of Technology (Lausanne). He then acted for many years as an editor of various academic journals, in particular *Defect and Diffusion Forum*. In recent years he has specialized in writing monographs which introduce readers to the most rapidly developing ideas in the fields of engineering, metallurgy and solid-state physics. He is co-author of the widely-cited student textbook, *Fundamentals of Solidification*, a new (5th fully-revised) edition of which is soon to appear. Google Scholar credits him with 8489 citations and a lifetime h-index of 14.

References

[1] Jansen, M., Angewandte Chemie, 41[20] 2002, 3746-3766. https://doi.org/10.1002/1521-3773(20021018)41:20<3746::AID-ANIE3746>3.0.CO;2-2

[2] Dshemuchadse, J., Steurer, W., Inorganic Chemistry, 54[3] 2015, 1120-1128. https://doi.org/10.1021/ic5024482

[3] King-Hele, D.G., Quarterly Journal of the Royal Astronomical Society, 13, 1972, 374-395.

[4] Tebay, S., Philosophical Magazine, 15, 1858, 206-212. https://doi.org/10.1080/14786445808642466

[5] Gamow, G., Proceedings of the National Academy of Sciences of the USA, 50, 1968, 313-318. https://doi.org/10.1073/pnas.59.2.313

[6] Lunn, A.C., Physical Review, 20[1] 1922, 1-14. https://doi.org/10.1103/PhysRev.20.1

[7] Batten, A.H., Quarterly Journal of the Royal Astronomical Society, 35, 1994, 249-270. https://doi.org/10.1093/mnras/270.1.35

[8] Sylvester, J.J., American Journal of Mathematics, 1, 1878, 64-125. https://doi.org/10.2307/2369436

[9] Crick, F.H.C., Griffith, J.S., Orgel, L.E., Proceedings of the National Academy of Sciences of the USA, 43, 1957, 416-421. https://doi.org/10.1073/pnas.43.5.416

[10] Svedberg, T., Proceedings of the Royal Society B, 127, 1939, 40-56.

[11] Lindgard, P.A., Bohr, H., Physical Review Letters, 77[4] 1996, 779-782. https://doi.org/10.1103/PhysRevLett.77.779

[12] Ray, A.K., Physical Review B, 48[19] 1993, 14702-14705. https://doi.org/10.1103/PhysRevB.48.14702

[13] Hamilton, J.C., Physical Review Letters, 77[5] 1996, 885-888. https://doi.org/10.1103/PhysRevLett.77.885

[14] Benfey, O.T., Journal of Chemical Education, 29[2] 1952, 78-81. https://doi.org/10.1021/ed029p78

[15] Restrepo, G., Pachón, L., Foundations of Chemistry, 9[2] 2007, 189-214. https://doi.org/10.1007/s10698-006-9026-6

[16] Dobereiner, J.W., Poggendorff's Annalen, 15, 1829, 301. https://doi.org/10.1002/andp.18290910217

[17] De Chancourtois, A.E.B., Comptes Rendus, 54, 1862, 757.

[18] Newlands, J.A.R., Chemical News, 10, 1864, 59.

[19] Cooke, J.P., American Journal of Science, 2[17] 1854, 387.

[20] Meyer, L., Annalen, VII, 1870, 354.

[21] Mazurs, E.G., Graphic Representations of the Periodic System during One Hundred Years, University of Alabama Press, 1974.

[22] Mendeleev, D., Annalen, VIII, 1871, 133.

[23] Villars, P., Journal of the Less-Common Metals, 92[2] 1983, 215-238. https://doi.org/10.1016/0022-5088(83)90489-7

[24] Pettifor, D., Journal of Physics C, 19, 1986, 285-313. https://doi.org/10.1088/0022-3719/19/3/002

[25] Pettifor, D.G., Journal of the Less-Common Metals, 114[1] 1985, 7-15. https://doi.org/10.1016/0022-5088(85)90384-4

[26] Pettifor, D.G., Journal of Physics C, 19[3] 1986, 285-313. https://doi.org/10.1088/0022-3719/19/3/002

[27] Pettifor, D.G., Physica B+C, 149[1-3] 1988, 3-10. https://doi.org/10.1016/0378-4363(88)90210-0

[28] Akdeniz, Z., Tosi, M.P., Journal of Physics - Condensed Matter, 1[13] 1989, 2381-2394. https://doi.org/10.1088/0953-8984/1/13/011

[29] Tosi, M.P., Pastore, G., Saboungi, M.L., Price, D.L., Physica Scripta, T39, 1991, 367-371. https://doi.org/10.1088/0031-8949/1991/T39/058

[30] Saboungi, M.L., Price, D.L., Scamehorn, C., Tosi, M.P., EPL, 15[3] 1991, 283-288. https://doi.org/10.1209/0295-5075/15/3/009

[31] Ramos, C., Saragovi, C., Granovsky, M.S., Journal of Nuclear Materials, 366[1-2] 2007, 198-205. https://doi.org/10.1016/j.jnucmat.2007.01.216

[32] Ohta, Y., Pettifor, D.G., Journal of Physics: Condensed Matter, 2[41] 1990, 8189-8194. https://doi.org/10.1088/0953-8984/2/41/006

[33] Kassem, M.A., Scripta Metallurgica et Materiala, 32[8] 1995, 1191-1196. https://doi.org/10.1016/0956-716X(95)00124-E

[34] Kiselyova, N.N., Stolyarenko, A.V., Senko, O.V., Dokukin, A.A., Russian Metallurgy, 2013[5] 2013, 381-388. https://doi.org/10.1134/S0036029513050091

[35] Pettifor, D.G., Journal of the Chemical Society, Faraday Transactions, 86[8] 1990, 1209-1213. https://doi.org/10.1039/ft9908601209

Materials Research Forum LLC
https://doi.org/10.21741/9781644902011

[36] Pettifor, D.G., Materials Science and Technology, 4[8] 1988, 675-691. https://doi.org/10.1179/mst.1988.4.8.675

[37] Pettifor, D.G., Journal of Physics Condensed Matter, 15[25] 2003, V13-V16. https://doi.org/10.1088/0953-8984/15/25/402

[38] Biswas, T., Ranganathan, S., Annales de Chimie - Science des Materiaux, 31[6] 2006, 649-656. https://doi.org/10.3166/acsm.31.649-656

[39] Ranganathan, S., Inoue, A., Acta Materialia, 54[14] 2006, 3647-3656. https://doi.org/10.1016/j.actamat.2006.01.041

[40] Vvedensky, D.D., Eberhart, M.E., Philosophical Magazine Letters, 55[4] 1987, 157-161. https://doi.org/10.1080/09500838708207549

[41] Morgan, D., Rodgers, J., Ceder, G., Journal of Physics - Condensed Matter, 15[25] 2003, 4361-4369. https://doi.org/10.1088/0953-8984/15/25/307

[42] Matysik, P., Czujko, T., Varin, R.A., International Journal of Hydrogen Energy, 39[1] 2014, 398-405. https://doi.org/10.1016/j.ijhydene.2013.10.104

[43] Munroe, P.R., Scripta Metallurgica et Materiala, 27[10] 1992, 1373-1378. https://doi.org/10.1016/0956-716X(92)90086-T

[44] Basu, J., Ranganathan, S., Intermetallics, 17[3] 2009, 128-135. https://doi.org/10.1016/j.intermet.2008.10.006

[45] Kawamura, T., Tachi, R., Inamura, T., Hosoda, H., Wakashima, K., Hamada, K., Miyazaki, S., Materials Science and Engineering A, 438-440, 2006, 383-386. https://doi.org/10.1016/j.msea.2006.01.123

[46] Hosoda, H., Wakashima, K., Miyazaki, S., Inoue, K., Ranganathan, S., Inoue, A., Materials Research Society Symposium Proceedings, 842, 2005, 353-358. https://doi.org/10.1557/PROC-842-S3.4

[47] Pettifor, D.G., Journal of Phase Equilibria, 17[5] 1996, 384-395. https://doi.org/10.1007/BF02667628

[48] Glawe, H., Sanna, A., Gross, E.K.U., Marques, M.A.L., New Journal of Physics, 18, 2016, 093011. https://doi.org/10.1088/1367-2630/18/9/093011

[49] Kiselyova, N.N., Senko, O.V., Kropotov, D.A., Dokukin, A.A., Russian Metallurgy, 2012[7], 2012, 644-653. https://doi.org/10.1134/S0036029512070087

[50] Zhang, R.F., Kong, X.F., Wang, H.T., Zhang, S.H., Legut, D., Sheng, S.H., Srinivasan, S., Rajan, K., Germann, T.C., Scientific Reports, 7[1] 2017, 9577. https://doi.org/10.1038/s41598-017-09704-1

[51] Pettifor, D.G., Solid State Physics - Advances in Research and Applications, 40[C] 1987, 43-92. https://doi.org/10.1016/S0081-1947(08)60690-6

[52] Ghiringhelli, L.M., Vybiral, J., Levchenko, S.V., Draxl, C., Scheffler, M., Physical Review Letters, 114[10] 2015, 105503. https://doi.org/10.1103/PhysRevLett.114.105503

[53] Ghiringhelli, L.M., Vybiral, J., Ahmetcik, E., Ouyang, R., Levchenko, S.V., Draxl, C., Scheffler, M., New Journal of Physics, 19[2] 2017, 023017. https://doi.org/10.1088/1367-2630/aa57bf

[54] Ouyang, R., Curtarolo, S., Ahmetcik, E., Scheffler, M., Ghiringhelli, L.M., Physical Review Materials, 2[8] 2018, 083802. https://doi.org/10.1103/PhysRevMaterials.2.083802

[55] Ferreira, L.G., Wei, S.H., Zunger, A., Physical Review B, 40[5] 1989, 3197-3231. https://doi.org/10.1103/PhysRevB.40.3197

[56] Le, D.H., Colinet, C., Hicter, P., Pasturel, A., Journal of Physics - Condensed Matter, 3[50] 1991, 9965-9974. https://doi.org/10.1088/0953-8984/3/50/002

[57] Colinet, C., Pasturel, A., Physica B, 192[3] 1993, 238-246. https://doi.org/10.1016/0921-4526(93)90026-3

[58] Sluiter, M.H.F., Colinet, C., Pasturel, A., Physical Review B, 73[17] 2006, 174204. https://doi.org/10.1103/PhysRevB.73.174204

[59] Sluiter, M.H.F., Acta Materialia, 55[11] 2007, 3707-3718. https://doi.org/10.1016/j.actamat.2007.02.016

[60] Valle, M., Oganov, A.R., IEEE Symposium on Visual Analytics Science and Technology, Proceedings, 4677351, 2008, 11-18.

[61] Levy, O., Hart, G.L.W., Curtarolo, S., Journal of the American Chemical Society, 132[13] 2010, 4830-4833. https://doi.org/10.1021/ja9105623

[62] Pettifor, D.G., Podloucky, R., Physical Review Letters, 53[11] 1984, 1080-1083. https://doi.org/10.1103/PhysRevLett.53.1080

[63] Pettifor, D.G., Podlouckyt, R., Journal of Physics C, 19[3] 1986, 315-330. https://doi.org/10.1088/0022-3719/19/3/003

[64] Pettifor, D.G., Physical Review Letters, 63[22] 1989, 2480-2483. https://doi.org/10.1103/PhysRevLett.63.2480

[65] Shah, M., Pettifor, D.G., Journal of Alloys and Compounds, 197[2] 1993, 145-152. https://doi.org/10.1016/0925-8388(93)90037-N

[66] Cressoni, J.C., Pettifor, D.G., Journal of Physics - Condensed Matter, 3[5] 1991, 495-511.

https://doi.org/10.1088/0953-8984/3/5/001

[67] Alinaghian, P., Gumbsch, P., Skinner, A.J., Pettifor, D.G., Journal of Physics - Condensed Matter, 5[32] 1993, 5795-5810. https://doi.org/10.1088/0953-8984/5/32/010

[68] Aoki, M., Pettifor, D.G., Materials Science and Engineering A, 176[1-2] 1994, 19-24. https://doi.org/10.1016/0921-5093(94)90954-7

[69] Pettifor, D.G., Aoki, M., Gumbsch, P., Horsfield, A.P., Nguyen-Manh, D., Vitek, V., Materials Science and Engineering A, 192-193[1] 1995, 24-30. https://doi.org/10.1016/0921-5093(94)03223-8

[70] Tan, K.E., Horsfield, A.P., Nguyen-Manh, D., Pettifor, D.G., Sutton, A.P., Physical Review Letters, 76[1] 1996, 90-93. https://doi.org/10.1103/PhysRevLett.76.90

[71] Ferrari, A., Schröder, M., Lysogorskiy, Y., Rogal, J., Mrovec, M., Drautz, R., Modelling and Simulation in Materials Science and Engineering, 27[8] 2019, 085008. https://doi.org/10.1088/1361-651X/ab471d

[72] Pettifor, D.G., Finnis, M.W., Nguyen-Manh, D., Murdick, D.A., Zhou, X.W., Wadley, H.N.G., Materials Science and Engineering A, 365[1-2] 2004, 2-13. https://doi.org/10.1016/j.msea.2003.09.001

[73] Pettifor, D.G., Oleynik, I.I., Progress in Materials Science, 49[3-4] 2004, 285-312. https://doi.org/10.1016/S0079-6425(03)00024-0

[74] Drautz, R., Murdick, D.A., Nguyen-Manh, D., Zhou, X., Wadley, H.N.G., Pettifor, D.G., Physical Review B, 72[14] 2005, 144105. https://doi.org/10.1103/PhysRevB.72.144105

[75] Hammerschmidt, T., Drautz, R., Pettifor, D.G., International Journal of Materials Research, 100[11] 2009, 1479-1487. https://doi.org/10.3139/146.110207

[76] Brutti, S., Nguyen-Manh, D., Pettifor, D.G., Journal of Alloys and Compounds, 457[1-2] 2008, 29-35. https://doi.org/10.1016/j.jallcom.2007.03.023

[77] Oganov, A.R., Valle, M., Journal of Chemical Physics, 130[10] 2009, 104504. https://doi.org/10.1063/1.3079326

[78] Seiser, B., Drautz, R., Pettifor, D.G., Acta Materialia, 59[2] 2011, 749-763. https://doi.org/10.1016/j.actamat.2010.10.013

[79] Seiser, B., Hammerschmidt, T., Kolmogorov, A.N., Drautz, R., Pettifor, D.G., Physical Review B, 83[22] 2011, 224116. https://doi.org/10.1103/PhysRevB.83.224116

[80] Pettifor, D.G., Seiser, B., Margine, E.R., Kolmogorov, A.N., Drautz, R., Philosophical Magazine, 93[28-30] 2013, 3907-3924. https://doi.org/10.1080/14786435.2013.771824

[81] Chen, Y., Kolmogorov, A.N., Pettifor, D.G., Shang, J.X., Zhang, Y., Physical Review B, 82[18] 2010, 184104. https://doi.org/10.1103/PhysRevB.82.184104

[82] Drain, J.F., Drautz, R., Pettifor, D.G., Physical Review B, 89[13] 2014, 134102. https://doi.org/10.1103/PhysRevB.89.134102

[83] Jenke, J., Subramanyam, A.P.A., Densow, M., Hammerschmidt, T., Pettifor, D.G., Drautz, R., Physical Review B, 98[14] 2018, 144102. https://doi.org/10.1103/PhysRevB.98.144102

[84] Watson, R.E., Bennett, L.H., Acta Metallurgica, 30[10] 1982, 1941-1955. https://doi.org/10.1016/0001-6160(82)90034-7

[85] Kuznetsov, V.N., Zhmurko, G.P., Sokolovskaya, E.M., Journal of the Less-Common Metals, 163[1] 1990, 1-8. https://doi.org/10.1016/0022-5088(90)90080-4

[86] Bieber, A., Gautier, F., Acta Metallurgica, 34[12] 1986, 2291-2309. https://doi.org/10.1016/0001-6160(86)90133-1

[87] Makino, Y., Intermetallics, 2[1] 1994, 55-66. https://doi.org/10.1016/0966-9795(94)90051-5

[88] Hoistad, L.M., Inorganic Chemistry, 34[10] 1995, 2711-2717. https://doi.org/10.1021/ic00114a033

[89] Chen, Y., Iwata, S., Liu, J., Villars, P., Rodgers, J., Modelling and Simulation in Materials Science and Engineering, 4[4] 1996, 335-348. https://doi.org/10.1088/0965-0393/4/4/001

[90] Harada, Y., Morinaga, M., Ito, A., Sugita, Y., Journal of Alloys and Compounds, 236[1-2] 1996, 92-101. https://doi.org/10.1016/0925-8388(95)02173-6

[91] Yi, D., Lai, Z., Li, C., Akselsen, O.M., Ulvensoen, J.H., Metallurgical and Materials Transactions A, 29[1] 1998, 119-129. https://doi.org/10.1007/s11661-998-0164-4

[92] Harada, Y., Morinaga, M., Saito, J.I., Takagi, Y., Journal of Physics - Condensed Matter, 9[38] 1997, 8011-8030. https://doi.org/10.1088/0953-8984/9/38/008

[93] Wheeler, R., Vasudevan, V.K., Fraser, H.L., Philosophical Magazine Letters, 62[3] 1990, 143-151. https://doi.org/10.1080/09500839008215051

[94] Eberhart, M.E., Kumar, K.S., Maclaren, J.M., Philosophical Magazine B, 61[6] 1990, 943-956. https://doi.org/10.1080/13642819008207854

[95] Eberhart, M.E., Clougherty, D.P., Maclaren, J.M., Philosophical Magazine B, 68[4] 1993, 455-464. https://doi.org/10.1080/13642819308217927

[96] Eberhart, M.E., Clougherty, D.P., MacLaren, J.M., Journal of Materials Research, 8[3] 1993, 438-448. https://doi.org/10.1557/JMR.1993.0438

[97] Eberhart, M.E., Vvedensky, D.D., Physical Review B, 37[14] 1988, 8488-8490. https://doi.org/10.1103/PhysRevB.37.8488

[98] Kleinke, H., Harbrecht, B., Zeitschrift fuer Anorganische und Allgemeine Chemie, 626[9] 2000, 1851-1853. https://doi.org/10.1002/1521-3749(200009)626:9<1851::AID-ZAAC1851>3.0.CO;2-#

[99] Derakhshan, S., Dashjav, E., Kleinke, H., Materials Research Society Symposium - Proceedings, 755, 2003, 341-346. https://doi.org/10.1557/PROC-755-DD8.6

[100] Lee, C.S., Dashjav, E., Kleinke, H., Chemistry of Materials, 13[11] 2001, 4053-4057. https://doi.org/10.1021/cm010433g

[101] Fornasini, M.L., Iandelli, A., Merlo, F., Pani, M., Intermetallics, 8[3] 2000, 239-246. https://doi.org/10.1016/S0966-9795(99)00111-9

[102] He, C., Li, G., Luo, Y., Li, Y., Rare Metals, 21[1] 2002, 28-35.

[103] Villars, P., Cenzual, K., Daams, J., Chen, Y., Iwata, S., Journal of Alloys and Compounds, 367[1-2] 2004, 167-175. https://doi.org/10.1016/j.jallcom.2003.08.060

[104] Derakhshan, S., Dashjav, E., Kleinke, H., European Journal of Inorganic Chemistry, 2004[6] 2004, 1183-1189. https://doi.org/10.1002/ejic.200300471

[105] Morinaga, M., Yukawa, H., Advanced Engineering Materials, 3[6] 2001, 381-385. https://doi.org/10.1002/1527-2648(200106)3:6<381::AID-ADEM381>3.0.CO;2-U

[106] Clark, P.M, Lee, S., Fredrickson, D.C., Journal of Solid State Chemistry, 178[4] 2005, 1269-1283. https://doi.org/10.1016/j.jssc.2004.12.044

[107] Levy, O., Hart, G.L.W., Curtarolo, S., Physical Review B, 81[17] 2010, 174106. https://doi.org/10.1103/PhysRevB.81.174106

[108] Kong, C.S., Villars, P., Iwata, S., Rajan, K., Computational Science and Discovery, 5[1] 2012, 015004. https://doi.org/10.1088/1749-4699/5/1/015004

[109] Kvashnin, A.G., Allahyari, Z., Oganov, A.R., Journal of Applied Physics, 126[4] 2019, 040901. https://doi.org/10.1063/1.5109782

[110] Zhang, Y., Mao, Z., Han, Q., Li, Y., Li, M., Du, S., Chai, Z., Huang, Q., Materialia, 12, 2020, 100810. https://doi.org/10.1016/j.mtla.2020.100810

[111] Villars, P., Phillips, J.C., Rabe, K.M., Brown, I.D., Ferroelectrics, 130[1] 1992, 129-135. https://doi.org/10.1080/00150199208019539

[112] Rabe, K.M., Phillips, J.C., Villars, P., Brown, I.D., Physical Review B, 45[14] 1992, 7650-7676. https://doi.org/10.1103/PhysRevB.45.7650

[113] Martin, S., Kirby, M., Miranda, R., Engineering Applications of Artificial Intelligence, 13[5], 2000, 513-520. https://doi.org/10.1016/S0952-1976(00)00030-0

[114] Curtarolo, S., Morgan, D., Persson, K., Rodgers, J., Ceder, G., Physical Review Letters, 91[13] 2003, 135503. https://doi.org/10.1103/PhysRevLett.91.135503

[115] Ceder, G., Morgan, D., Fischer, C., Tibbetts, K., Curtarolo, S., MRS Bulletin, 31[12] 2006, 981-985. https://doi.org/10.1557/mrs2006.224

[116] Le Page, Y., MRS Bulletin, 31[12] 2006, 991-994. https://doi.org/10.1557/mrs2006.226

[117] Fischer, C.C., Tibbetts, K.J., Morgan, D., Ceder, G., Nature Materials, 5[8] 2006, 641-646. https://doi.org/10.1038/nmat1691

[118] Balachandran, P.V., Rajan, K., Acta Crystallographica B, 68[1] 2012, 24-33. https://doi.org/10.1107/S0108768111054061

[119] Kong, C.S., Luo, W., Arapan, S., Villars, P., Iwata, S., Ahuja, R., Rajan, K., Journal of Chemical Information and Modeling, 52[7] 2012, 1812-1820. https://doi.org/10.1021/ci200628z

[120] Yang, L., Ceder, G., Physical Review B, 88[22] 2013, 224107. https://doi.org/10.1103/PhysRevB.88.224107

[121] Curtarolo, S., Hart, G.L.W., Nardelli, M.B., Mingo, N., Sanvito, S., Levy, O., Nature Materials, 12[3] 2013, 191-201. https://doi.org/10.1038/nmat3568

[122] Meredig, B., Wolverton, C., Chemistry of Materials, 26[6] 2014, 1985-1991. https://doi.org/10.1021/cm403727z

[123] Hart, G.L.W., Curtarolo, S., Massalski, T.B., Levy, O., Physical Review X, 3[4] 2014, e041035. https://doi.org/10.1103/PhysRevX.3.041035

[124] Carrete, J., Li, W., Mingo, N., Wang, S., Curtarolo, S., Physical Review X, 4[1] 2014, 011019. https://doi.org/10.1103/PhysRevX.4.011019

[125] Carrete, J., Mingo, N., Wang, S., Curtarolo, S., Advanced Functional Materials, 24[47] 2014 7427-7432. https://doi.org/10.1002/adfm.201401201

[126] Oliynyk, A.O., Antono, E., Sparks, T.D., Ghadbeigi, L., Gaultois, M.W., Meredig, B., Mar, A., Chemistry of Materials, 28[20] 2016, 7324-7331. https://doi.org/10.1021/acs.chemmater.6b02724

[127] Zheng, X., Zheng, P., Zhang, R.Z., Chemical Science, 9[44] 2018, 8426-8432. https://doi.org/10.1039/C8SC02648C

[128] Fang, T., Zhao, X., Zhu, T., Materials, 11[5] 2018, 847. https://doi.org/10.3390/ma11050847

[129] Angsten, T., Mayeshiba, T., Wu, H., Morgan, D., New Journal of Physics, 16, 2014, 015018. https://doi.org/10.1088/1367-2630/16/1/015018

[130] Dshemuchadse, J., Steurer, W., Acta Crystallographica A, 71, 2015, 335-345. https://doi.org/10.1107/S2053273315004064

[131] Bialon, A.F., Hammerschmidt, T., Drautz, R., Chemistry of Materials, 28[8] 2016, 2550-2556. https://doi.org/10.1021/acs.chemmater.5b04299

[132] Engelkemier, J., Green, L.M., McDougald, R.N., McCandless, G.T., Chan, J.Y., Fredrickson, D.C., Crystal Growth and Design, 16[9] 2016, 5349-5358. https://doi.org/10.1021/acs.cgd.6b00855

[133] Oliynyk, A.O., Adutwum, L.A., Harynuk, J.J., Mar, A., Chemistry of Materials, 28[18] 2016, 6672-6681. https://doi.org/10.1021/acs.chemmater.6b02905

[134] Jain, A., Hautier, G., Ong, S.P., Persson, K., Journal of Materials Research, 31[8] 2016, 977-994. https://doi.org/10.1557/jmr.2016.80

[135] Kiselyova, N.N., Dudarev, V.A., Korzhuyev, M.A., Inorganic Material - Applied Research, 7[1] 2016, 34-39. https://doi.org/10.1134/S2075113316010093

[136] Xian, Y., Zheng, H., Zhai, Q., Luo, Z., Computational Materials Science, 125, 2016, 1-7. https://doi.org/10.1016/j.commatsci.2016.08.023

[137] Nyshadham, C., Oses, C., Hansen, J.E., Takeuchi, I., Curtarolo, S., Hart, G.L.W., Acta Materialia, 122, 2017, 438-447. https://doi.org/10.1016/j.actamat.2016.09.017

[138] Levy, O., Chepulskii, R.V., Hart, G.L.W., Curtarolo, S., Journal of the American Chemical Society, 132[2] 2010, 833-837. https://doi.org/10.1021/ja908879y

[139] Levy, O., Hart, G.L.W., Curtarolo, S., Acta Materialia, 58[8] 2010, 2887-2897. https://doi.org/10.1016/j.actamat.2010.01.017

[140] Levy, O., Jahnátek, M., Chepulskii, R.V., Hart, G.L.W., Curtarolo, S., Journal of the American Chemical Society, 133[1] 2011, 158-163. https://doi.org/10.1021/ja1091672

[141] Jahnátek, M., Levy, O., Hart, G.L.W., Nelson, L.J., Chepulskii, R.V., Xue, J., Curtarolo, S., Physical Review B, 84[21] 2011, 214110. https://doi.org/10.1103/PhysRevB.84.214110

[142] Levy, O., Xue, J., Wang, S., Hart, G.L.W., Curtarolo, S., Physical Review B, 85[1] 2012, 012201. https://doi.org/10.1103/PhysRevB.85.012201

[143] Goldsmith, B.R., Boley, M., Vreeken, J., Scheffler, M., Ghiringhelli, L.M., New Journal of Physics, 19[1] 2017, 013031. https://doi.org/10.1088/1367-2630/aa57c2

[144] Schmidt, J., Shi, J., Borlido, P., Chen, L., Botti, S., Marques, M.A.L., Chemistry of

Materials Research Forum LLC
https://doi.org/10.21741/9781644902011

Materials, 29[12] 2017, 5090-5103. https://doi.org/10.1021/acs.chemmater.7b00156

[145] Takahashi, K., Takahashi, L., Journal of Physical Chemistry Letters, 10[2] 2019, 283-288. https://doi.org/10.1021/acs.jpclett.8b03527

[146] Zhao, Y., Cui, Y., Xiong, Z., Jin, J., Liu, Z., Dong, R., Hu, J., ACS Omega, 5[7] 2020, 3596-3606. https://doi.org/10.1021/acsomega.9b04012

[147] Allahyari, Z., Oganov, A.R., NPJ Computational Materials, 6[1] 2020, 55. https://doi.org/10.1038/s41524-020-0322-9

[148] Hammerschmidt, T., Bialon, A.F., Drautz, R., Modelling and Simulation in Materials Science and Engineering, 25[7] 2017, 074002. https://doi.org/10.1088/1361-651X/aa83c3

[149] Oliynyk, A.O., Adutwum, L.A., Rudyk, B.W., Pisavadia, H., Lotfi, S., Hlukhyy, V., Harynuk, J.J., Mar, A., Brgoch, J., Journal of the American Chemical Society, 13[49] 2017, 17870-17881. https://doi.org/10.1021/jacs.7b08460

[150] Li, L., You, Y., Hu, S., Shi, Y., Zhao, G., Chen, C., Wang, Y., Stroppa, A., Ren, W., Applied Physics Letters, 114[8] 2019, 083102. https://doi.org/10.1063/1.5045512

[151] Narasimhan, S., APL Materials, 8[4] 2020, 040903. https://doi.org/10.1063/5.0003256

[152] Hu, J., Cao, Z., Dan, Y., Niu, C., Li, X., Qian, S., Journal of South China University of Technology, 47[5] 2019, 48-55.

[153] Ferris, K.F., Webb-Robertson, B.J.M., Jones, D.M., Proceedings of SPIE, 6403, 2007, 64032A.

[154] Pilania, G., Gubernatis, J.E., Lookman, T., Scientific Reports, 5, 2015, 17504. https://doi.org/10.1038/srep17504

[155] Pilania, G., Gubernatis, J.E., Lookman, T., Physical Review B, 91[21] 2015, 214302. https://doi.org/10.1103/PhysRevB.91.214302

[156] Zhang, H., Guo, Z., Hu, H., Zhou, G., Liu, Q., Xu, Y., Qian, Q., Dai, D., Modelling and Simulation in Materials Science and Engineering, 28[3] 2020, 035002.

[157] Gopakumar, A.M., Balachandran, P.V., Xue, D., Gubernatis, J.E., Lookman, T., Scientific Reports, 8[1] 2018, 3738. https://doi.org/10.1038/s41598-018-21936-3

[158] Rickman, J.M., Lookman, T., Kalinin, S.V., Acta Materialia, 168, 2019, 473-510. https://doi.org/10.1016/j.actamat.2019.01.051

[159] Cardinale, A.M., Parodi, N., Macciò, D., Journal of Phase Equilibria and Diffusion, 39[6] 2018, 908-915. https://doi.org/10.1007/s11669-018-0692-6

[160] Hever, A., Oses, C., Curtarolo, S., Levy, O., Natan, A., Inorganic Chemistry, 57[2] 2018,

653-667. https://doi.org/10.1021/acs.inorgchem.7b02462

[161] Konar, B., Kim, J., Jung, I.H., Journal of Phase Equilibria and Diffusion, 38[4] 2017, 509-542. https://doi.org/10.1007/s11669-017-0546-7

[162] Davies, D.W., Butler, K.T., Jackson, A.J., Morris, A., Frost, J.M., Skelton, J.M., Walsh, A., Chem, 1[4] 2016, 617-627. https://doi.org/10.1016/j.chempr.2016.09.010

[163] Song, Y., Dai, J.H., Yang, R., Advanced Engineering Materials and Modeling, 2016, 203-228. https://doi.org/10.1002/9781119242567.ch7

[164] Lowe, D.H., IOP Conference Series - Materials Science and Engineering, 117[1] 2016, 012026. https://doi.org/10.1088/1757-899X/117/1/012026

[165] Han, C.S., Lee, S.M., Asian Journal of Chemistry, 28[6] 2016, 1215-1217. https://doi.org/10.14233/ajchem.2016.19626

[166] Hammerschmidt, T., Ladines, A.N., Kossmann, J., Drautz, R., Crystals, 6[2] 2016, 18. https://doi.org/10.3390/cryst6020018

[167] Schön, C.G., Gil Rebaza, A.V., Fernández, V.I., Eleno, L.T.F., Gonzales-Ormeño, P.G., Errico, L.A., Petrilli, H.M., Journal of Alloys and Compounds, 688, 2016, 337-341. https://doi.org/10.1016/j.jallcom.2016.07.205

[168] Gjoka, M., Psycharis, V., Devlin, E., Niarchos, D., Hadjipanayis, G., Journal of Alloys and Compounds, 687, 2016, 240-245. https://doi.org/10.1016/j.jallcom.2016.06.098

[169] Medasani, B., Haranczyk, M., Canning, A., Asta, M., Computational Materials Science, 101, 2015, 96-107. https://doi.org/10.1016/j.commatsci.2015.01.018

[170] Zurek, E., Grochala, W., Physical Chemistry Chemical Physics, 17[5] 2015, 2917-2934. https://doi.org/10.1039/C4CP04445B

[171] Ning, J., Zhang, X., Qin, J., Liu, Y., Ma, M., Liu, R., Journal of Alloys and Compounds, 618, 2015, 73-77. https://doi.org/10.1016/j.jallcom.2014.08.168

[172] Coetzee, S.H., Cornish, L.A., Witcomb, M.J., Jain, P.K., Journal of Phase Equilibria and Diffusion, 36[2] 2015, 149-168. https://doi.org/10.1007/s11669-015-0368-4

[173] Cardinale, A.M., Macciò, D., Calphad, 46, 2014, 220-225. https://doi.org/10.1016/j.calphad.2014.05.001

[174] Soliman, S., Journal of Alloys and Compounds, 571, 2013, 69-74. https://doi.org/10.1016/j.jallcom.2013.03.035

[175] Von Lilienfeld, O.A., International Journal of Quantum Chemistry, 113[12] 2013, 1676-1689. https://doi.org/10.1002/qua.24375

[176] Nelson, L.J., Hart, G.L.W., Zhou, F., Ozoliņš, V., Physical Review B, 87[3] 2013, 035125. https://doi.org/10.1103/PhysRevB.87.035125

[177] Svane, A., Albers, R.C., Christensen, N.E., Van Schilfgaarde, M., Chantis, A.N., Zhu, J.X., Physical Review B, 87[4] 2013, 045109. https://doi.org/10.1103/PhysRevB.87.045109

[178] Tegner, B.E., Zhu, L., Ackland, G.J., Physical Review B, 85[21] 2012, 214106. https://doi.org/10.1103/PhysRevB.85.214106

[179] Rajasekharan, T., Seshubai, V., Acta Crystallographica A, 68[1] 2012, 156-165. https://doi.org/10.1107/S0108767311044151

[180] McEniry, E.J., Madsen, G.K.H., Drain, J.F., Drautz, R., Journal of Physics - Condensed Matter, 23[27] 2011, 276004. https://doi.org/10.1088/0953-8984/23/27/276004

[181] Wang, Y.J., Wang, C.Y., Materials Research Society Symposium Proceedings, 1224, 2010, 169-177.

[182] Yin, Z.P., Pickett, W.E., Physical Review B, 82[15] 2010, 155202. https://doi.org/10.1103/PhysRevB.82.155202

[183] Feng, J., Hoffmann, R., Ashcroft, N.W., Journal of Chemical Physics, 132[11] 2010, 114106. https://doi.org/10.1063/1.3328198

[184] Simonovic, D., Sluiter, M.H.F., Physical Review B, 79[5] 2009, 054304. https://doi.org/10.1103/PhysRevB.79.054304

[185] Feng, J., Hennig, R.G., Ashcroft, N.W., Hoffmann, R., Nature, 451[7177] 2008, 445-448. https://doi.org/10.1038/nature06442

[186] Wu, Y., Hu, W., Materials Research and Advanced Techniques, 99[1] 2008, 42-49. https://doi.org/10.3139/146.101609

[187] Wu, Y., Hu, W., International Journal of Materials Research, 99[1] 2008, 42-49. https://doi.org/10.3139/146.101609

[188] Schwarz, W.H.E., Foundations of Chemistry, 9[2] 2007, 139-188. https://doi.org/10.1007/s10698-006-9020-z

[189] Bhatt, J., Ray, P.K., Murty, B.S., Transactions of the Indian Institute of Metals, 60[2-3] 2007, 323-330.

[190] Hart, G.L.W., Nature Materials, 6[12] 2007, 941-945. https://doi.org/10.1038/nmat2057

[191] Kolb, B., Müller, S., Botts, D.B., Hart, G.L.W., Physical Review B, 74[14] 2006, 144206. https://doi.org/10.1103/PhysRevB.74.144206

[192] Ruhnow, M., Philosophical Magazine, 85[17] 2005, 1865-1899.

Materials Research Forum LLC
https://doi.org/10.21741/9781644902011

https://doi.org/10.1080/14786430500098793

[193] Riani, P., Arrighi, L., Marazza, R., Mazzone, D., Zanicchi, G., Ferro, R., Intermetallics, 13[6] 2005, 669-680. https://doi.org/10.1016/j.intermet.2004.10.013

[194] Derakhshan, S., Assoud, A., Kleinke, K.M., Dashjav, E., Qiu, X., Billinge, S.J.L., Kleinke, H., Journal of the American Chemical Society, 126[26] 2004, 8295-8302. https://doi.org/10.1021/ja048262e

[195] Stein, F., Palm, M., Sauthoff, G., Intermetallics, 12[7-9] 2004, 713-720. https://doi.org/10.1016/j.intermet.2004.02.010

[196] Müller, S., Journal of Physics - Condensed Matter, 15[34] 2003, R1429-R1500. https://doi.org/10.1088/0953-8984/15/34/201

[197] Plieth, W., Surface and Coatings Technology, 169-170, 2003, 96-99. https://doi.org/10.1016/S0257-8972(03)00166-X

[198] Boettinger, W.J., Vaudin, M.D., Williams, M.E., Bendersky, L.A., Wagner, W.R., Journal of Electronic Materials, 32[6] 2003, 511-515. https://doi.org/10.1007/s11664-003-0135-x

[199] Tang, N.Y., Su, X., Yu, X., Materials Research and Advanced Techniques, 94[2] 2003, 116-121. https://doi.org/10.3139/146.030116

[200] Blank, H., Journal of Alloys and Compounds, 343[1-2] 2002, 90-107. https://doi.org/10.1016/S0925-8388(02)00122-6

[201] Ferro, R., Cacciamani, G., Calphad, 26[3] 2002, 439-458. https://doi.org/10.1016/S0364-5916(02)00056-1

[202] Granovsky, M.S., Canay, M., Lena, E., Arias, D., Journal of Nuclear Materials, 302[1] 2002, 1-8. https://doi.org/10.1016/S0022-3115(02)00718-3

[203] Ferro, R., Cacciamani, G., TMS Annual Meeting, 2002, 177-189.

[204] Studnitzky, T., Schmid-Fetzer, R., Materials Research and Advanced Techniques, 93[9] 2002, 894-903. https://doi.org/10.3139/146.020894

[205] Du, X.W., Zhu, J., Wang, B., Kim, Y.W., Philosophical Magazine A, 82[1] 2002, 39-50. https://doi.org/10.1080/01418610208239995

[206] Hirschl, R., Hafner, J., Jeanvoine, Y., Journal of Physics - Condensed Matter, 13[14] 2001, 3545-3572. https://doi.org/10.1088/0953-8984/13/14/324

[207] Batsanov, S.S., Inorganic Materials, 37[1] 2001, 23-30. https://doi.org/10.1023/A:1026773024407

[208] Tang, N.Y., Journal of Phase Equilibria, 21[1] 2000, 70-77.

https://doi.org/10.1361/105497100770340444

[209] Christensen, S.W., Thomas, N.W., Acta Crystallographica A, 55[5] 1999, 811-820.
https://doi.org/10.1107/S0108767399001622

[210] Häussermann, U., Svensson, C., Lidin, S., Journal of the American Chemical Society, 120[16] 1998, 3867-3880. https://doi.org/10.1021/ja973335y

[211] Krishnan, S., Ansell, S., Price, D.L., Journal of the American Ceramic Society, 81[7] 1998, 1967-1969. https://doi.org/10.1111/j.1151-2916.1998.tb02578.x

[212] Neumann, A., Nguyen-Manh, D., Physical Review B, 57[18] 1998, 11149-11157. https://doi.org/10.1103/PhysRevB.57.11149

[213] Muller, D.A., Batson, P.E., Physical Review B, 58[18] 1998, 11970-11981. https://doi.org/10.1103/PhysRevB.58.11970

[214] Saccone, A., Cacciamani, G., Macciò, D., Borzone, G., Ferro, R., Intermetallics, 6[3] 1998, 201-215. https://doi.org/10.1016/S0966-9795(97)00066-6

[215] Giovannini, M., Saccone, A., Flandorfer, H., Rogl, P., Ferro, R., Materials Research and Advanced Techniques, 88[5] 1997, 372-378.

[216] Häussermann, U., Simak, S.I., Abrikosov, I.A., Lidin, S., Chemistry - a European Journal, 3[6] 1997, 904-911. https://doi.org/10.1002/chem.19970030612

[217] Zhukov, V.P., Journal of Structural Chemistry, 38[3] 1997, 459-482. https://doi.org/10.1007/BF02763614

[218] Thomas, N.W., Acta Crystallographica B, 52[6] 1996, 939-953. https://doi.org/10.1107/S0108768196009202

[219] Häussermann, U., Wörle, M., Nesper, R., Journal of the American Chemical Society, 118[47] 1996, 11789-11797. https://doi.org/10.1021/ja961127k

[220] Lee, S., Annual Review of Physical Chemistry, 47, 1996, 397-419. https://doi.org/10.1146/annurev.physchem.47.1.397

[221] Lee, S., Hoistad, L., Journal of Alloys and Compounds, 229[1] 1995, 66-79. https://doi.org/10.1016/0925-8388(95)01687-2

[222] Burdett, J.K., Journal of Molecular Structure, 336[2-3] 1995, 115-136. https://doi.org/10.1016/0166-1280(94)04073-2

[223] Häussermann, U., Nesper, R., Journal of Alloys and Compounds, 218[2] 1995, 244-254. https://doi.org/10.1016/0925-8388(94)01388-8

[224] Pei, S., Massalski, T.B., Acta Metallurgica et Materialia, 43[6] 1995, 2385-2394.

https://doi.org/10.1016/0956-7151(94)00439-0

[225] Narasimhan, S., Davenport, J.W., Physical Review B, 51[1] 1995, 659-662. https://doi.org/10.1103/PhysRevB.51.659

[226] Burdett, J.K., Marians, C., Mitchell, J.F., Inorganic Chemistry, 33[9] 1994, 1848-1856. https://doi.org/10.1021/ic00087a020

[227] Ogawa, T., Journal of Alloys and Compounds, 203[C] 1994, 221-227. https://doi.org/10.1016/0925-8388(94)90739-0

[228] Lee, S., Foran, B., Journal of the American Chemical Society, 116[1] 1994, 154-161. https://doi.org/10.1021/ja00080a018

[229] Ferro, Delfino, S., Borzone, G., Saccone, A., Cacciamani, G., Journal of Phase Equilibria, 14[3] 1993, 273-279. https://doi.org/10.1007/BF02668224

[230] Hoistad, L.M., Lee, S., Pasternak, J., Journal of the American Chemical Society, 114[12] 1992, 4790-4796. https://doi.org/10.1021/ja00038a051

[231] Erbölükbas, A., Akdeniz, Z., Tosi, M.P., Il Nuovo Cimento D, 14[1] 1992, 87-97. https://doi.org/10.1007/BF02455347

[232] Lee, S., Rousseau, R., Wells, C., Physical Review B, 46[19] 1992, 12121-12131. https://doi.org/10.1103/PhysRevB.46.12121

[233] Yasuda, H., Takasugi, T., Koiwa, M., Acta Metallurgica et Materialia, 40[2] 1992, 381-387. https://doi.org/10.1016/0956-7151(92)90312-3

[234] Pastore, G., Akdeniz, Z., Tosi, M.P., Journal of Physics - Condensed Matter, 3[42] 1991, 8297-8304. https://doi.org/10.1088/0953-8984/3/42/024

[235] Le, D.H., Colinet, C., Hicter, P., Pasturel, A., Journal of Physics - Condensed Matter, 3[40] 1991, 7895-7906. https://doi.org/10.1088/0953-8984/3/40/011

[236] Akdeniz, Z., Tosij, M.P., Philosophical Magazine B, 64[2] 1991, 167-179. https://doi.org/10.1080/13642819108207612

[237] Akdeniz, Z., Tosi, M.P., Journal of Non-Crystalline Solids, 117-118[2] 1990, 642-645. https://doi.org/10.1016/0022-3093(90)90613-Q

[238] Ackland, G.J., Vitek, V., Physical Review B, 41[15] 1990, 10324-10333. https://doi.org/10.1103/PhysRevB.41.10324

[239] Kumar, K.S., International Materials Reviews, 35[1] 1990, 293-328. https://doi.org/10.1179/095066090790324037

[240] Cranshaw, T.E., Journal of Physics - Condensed Matter, 1[36] 1989, 6431-6440.

https://doi.org/10.1088/0953-8984/1/36/011

[241] Sarma, D.D., Speier, W., Zeller, R., Van Leuken, E., De Groot, R.A., Fuggle, J.C., Journal of Physics - Condensed Matter, 1[46] 1989, 9131-9139. https://doi.org/10.1088/0953-8984/1/46/007

[242] Guo, X.Q., Podloucky, R., Freeman, A.J., Physical Review B, 40[5] 1989, 2793-2800. https://doi.org/10.1103/PhysRevB.40.2793

[243] Paskowicz, W., Journal of Physics F, 18[8] 1988, 1761-1785. https://doi.org/10.1088/0305-4608/18/8/014

[244] Akdeniz, Z., Li, W., Tosi, M.P., EPL, 5[7] 1988, 613-617. https://doi.org/10.1209/0295-5075/5/7/007

[245] Lopez, J.M., Alonso, J.A., Gallego, L.J., Silbert, M., Physica B+C, 150[3] 1988, 369-377. https://doi.org/10.1016/0378-4363(88)90077-0

[246] Hara, S., Ohdomari, I., Physical Review B, 38[11] 1988, 7554-7557. https://doi.org/10.1103/PhysRevB.38.7554

[247] Andreoni, W., Galli, G., Physics and Chemistry of Minerals, 14[5] 1987, 389-395. https://doi.org/10.1007/BF00628814

[248] Andreoni, W., Galli, G., Physical Review Letters, 58[26] 1987, 2742-2745. https://doi.org/10.1103/PhysRevLett.58.2742

[249] Majewski, J.A., Vogl, P., Physical Review B, 35[18] 1987, 9666-9682. https://doi.org/10.1103/PhysRevB.35.9666

[250] Majewski, J.A., Vogl, P., Physical Review Letters, 57[11] 1986, 1366-1369. https://doi.org/10.1103/PhysRevLett.57.1366

[251] Andreoni, W., Galli, G., Tosi, M., Physical Review Letters, 55[17] 1985, 1734-1737. https://doi.org/10.1103/PhysRevLett.55.1734

[252] Enders, P., Physica Status Solidi B, 132[1] 1985, 165-172. https://doi.org/10.1002/pssb.2221320117

www.ingramcontent.com/pod-product-compliance
Lightning Source LLC
Chambersburg PA
CBHW071716210326
41597CB00017B/2500